Communications in Computer and Information Science 957

Commenced Publication in 2007
Founding and Former Series Editors:
Phoebe Chen, Alfredo Cuzzocrea, Xiaoyong Du, Orhun Kara, Ting Liu,
Dominik Ślęzak, and Xiaokang Yang

Editorial Board

More information about this series at http://www.springer.com/series/7899

Jun Zhao · Frank van Harmelen
Jie Tang · Xianpei Han
Quan Wang · Xianyong Li (Eds.)

Knowledge Graph and Semantic Computing

Knowledge Computing and Language Understanding

Third China Conference, CCKS 2018
Tianjin, China, August 14–17, 2018
Revised Selected Papers

 Springer

Editors
Jun Zhao
Institute of Automation
Chinese Academy of Sciences
Beijing, China

Frank van Harmelen
Department of Computer Science
Vrije Universiteit Amsterdam
Amsterdam, The Netherlands

Jie Tang
Tsinghua University
Beijing, China

Xianpei Han
Institute of Software
Chinese Academy of Sciences
Beijing, China

Quan Wang
Institute of Information Engineering
Chinese Academy of Sciences
Beijing, China

Xianyong Li
Xihua University
Chendu, China

ISSN 1865-0929   ISSN 1865-0937 (electronic)
Communications in Computer and Information Science
ISBN 978-981-13-3145-9   ISBN 978-981-13-3146-6   (eBook)
https://doi.org/10.1007/978-981-13-3146-6

Library of Congress Control Number: 2018962353

This Springer imprint is published by the registered company Springer Nature Singapore Pte Ltd.
The registered company address is: 152 Beach Road, #21-01/04 Gateway East, Singapore 189721, Singapore

Preface

This volume contains the papers presented at CCKS 2018: the China Conference on Knowledge Graph and Semantic Computing held during August 14–17, 2018, in Tianjin.

CCKS is organized by the Technical Committee on Language and Knowledge Computing of the Chinese Information Processing Society (CIPS). CCKS 2018 was the third edition of the conference series, where the first edition was held in Beijing in 2016, and the second edition held in Chengdu in 2017. The first edition, CCKS 2016, was the merger of two premier relevant forums held previously: the Chinese Knowledge Graph Symposium (KGS) and the Chinese Semantic Web and Web Science Conference (CSWS). KGS was held in Beijing in 2013, in Nanjing in 2014, and in Yichang in 2015. CSWS was first held in Beijing in 2006, and has been the main forum for research on the Semantic (Web) technologies in China for nearly ten years. CCKS covers wide research fields including knowledge graph, the Semantic Web, linked data, NLP, knowledge representation, graph databases etc. It aims to be the top forum on knowledge graph and semantic technologies for Chinese researchers and practitioners from academia, industry, and government.

The theme of this year was "Knowledge Computing and Language Understanding."

There were 101 submissions. Each submission was reviewed by at least two, and on average 3.1, Program Committee members. The committee decided to accept 29 full/short papers (including 15 papers written in English and 14 papers written in Chinese). The program also included four invited keynotes, six tutorials, four shared tasks, two industrial forums, and one panel. The CCKS volume contains revised versions of 12 full/short papers written in English. This year's invited talks were given by Prof. Bo Zhang from Tsinghua University, Prof. James A. Hendler from Rensselaer Polytechnic Institute, Dr. Hui Qiang from Alibaba, and Prof. Roberto Navigli from Sapienza University of Rome.

The hard work and close collaboration of a number of people contributed to the success of this conference. We would like to thank the members of the Organizing Committee and Program Committee for their support as well as the authors and participants, who are the primary reason of the success of this conference.

Finally, we appreciate the sponsorship from AISPEECH, LEGAL INTELLIGENCE, Meituan Dianping, WeBank, Alibaba Group, and GTCOM as platinum sponsors, Unisound, PING AN Healthcare Technology, yuri.ai, ChinaScope, and Alex & Masen as gold sponsors, PlantData, Credit Harmony Research,

Xiaomi Inc., and YIDUCLOUD as silver sponsors, and SoftBank Robotics and MEMECT as bronze sponsors.

September 2018   Jun Zhao
                                                            Frank van Harmelen
                                                                      Jie Tang
                                                                  Xianpei Han
                                                                   Quan Wang
                                                                 Xianyong Li

Organization

CCKS 2018 was organized by the Technical Committee on Language and Knowledge Computing of the Chinese Information Processing Society.

General Chairs

Jun Zhao   Institute of Automation, Chinese Academy of Sciences, China
Frank van Harmelen   Vrije Universiteit Amsterdam, The Netherlands

Program Committee Chairs

Jie Tang   Tsinghua University, China
Xianpei Han   Institute of Software, Chinese Academy of Sciences, China

Local Chairs

Yuzhi Zhang   Nankai University, China
Xiaowang Zhang   Tianjin University, China

Publicity Chairs

Guilin Qi   Southeast University, China
Bin Xu   Tsinghua University, China

Publication Chairs

Quan Wang   Institute of Information Engineering, Chinese Academy
     of Sciences, China
Xianyong Li   Xihua University, China

Tutorial Chairs

Huajun Chen   Zhejiang University, China
Xiaodan Zhu   Queen's University, Canada

Evaluation Chairs

Haofen Wang   Shenzhen Gowild Intelligent Technology Company, China
Lei Zou   Peking University, China

Top Conference Reviewing Chairs

Yanghua Xiao   Fudan University, China
Ming Liu   Harbin Institute of Technology, China
Qili Zhu   Shanghai Jiao Tong University, China

Poster/Demo Chairs

Tieyun Qian   Wuhan University, China
Wei Hu   Nanjing University, China

Sponsorship Chairs

Kang Liu   Institute of Automation, Chinese Academy of Sciences, China
Siwei Yu   Guizhou Medical University, China

Industry Track Chairs

Wei Zhang   Alibaba Group, China
Xipeng Qiu   Fudan University, China

Panel Chairs

Xiaoyan Zhu   Tsinghua University, China
Bing Qin   Harbin Institute of Technology, China

Meta Data Chair

Jie Bao   MEMECT, China

Website Chair

Yubo Chen   Institute of Automation, Chinese Academy of Sciences, China

Area Chairs

Representation Learning for Knowledge Graph
Quan Wang   Institute of Information Engineering, Chinese Academy
                          of Sciences, China
Yang Yang   Zhejiang University, China

Knowledge Graph Construction and Information Extraction
Longhua Qian   Soochow University, China
Huawei Shen   Institute of Computing Technology, Chinese Academy
                          of Sciences, China

Mining Heterogeneous Knowledge Graphs

Chuan Shi   Beijing University of Posts and Telecommunications, China
Lei Zou   Peking University, China

Knowledge Storage and Indexing

Yanghua Xiao   Fudan University, China
Qili Zhu   Shanghai Jiao Tong University, China

Language Understanding and Machine Reading

Yansong Feng   Peking University, China
Tong Xu   University of Science and Technology of China, China

Question Answering and Semantic Search

Minlie Huang   Tsinghua University, China
Daifeng Li   Sun Yat-sen University, China

Linked Data and Semantic Integration

Gong Cheng   Nanjing University, China
Zhichun Wang   Beijing Normal University, China

Knowledge Representation and Reasoning

Guilin Qi   Southeast University, China
Wei Hu   Nanjing University, China

Program Committee

Peng Bao   Beijing Jiaotong University, China
Huajun Chen   Zhejiang University, China
Penghe Chen   Beijing Normal University, China
Wenliang Chen   Soochow University, China
Yubo Chen   Institute of Automation, Chinese Academy of Sciences, China
Zhumin Chen   Shandong University, China
Gong Cheng   Nanjing University, China
Wanyun Cui   Shanghai University of Finance and Economics, China
Xinyu Dai   Nanjing University, China
Jianfeng Du   Guangdong University of Foreign Studies, China
Nan Duan   Microsoft Research Asia, China
Tianqing Fang   Zhejiang University, China
Yansong Feng   Peking University, China
Jibing Gong   Institute of Computing Technology, Chinese Academy
                         of Sciences, China
Jinguang Gu   Wuhan University of Science and Technology, China

Shu Guo	Institute of Information Engineering, Chinese Academy of Sciences, China
Yuhang Guo	Beijing Institute of Technology, China
Xianpei Han	Institute of Software, Chinese Academy of Sciences, China
Tianyong Hao	Guangdong University of Foreign Studies, China
Liang Hong	Wuhan University, China
Ling Hong	Wuhan University, China
Linmei Hu	Tsinghua University, China
Wei Hu	Nanjing University, China
Minlie Huang	Tsinghua University, China
Qiu Ji	Nanjing University of Posts and Telecommunications, China
Xiaolong Jin	Institute of Computing Technology, Chinese Academy of Sciences, China
Binyang Li	University of International Relations, China
Daifeng Li	Sun Yat-sen University, China
Huiying Li	Southeast University, China
Lishuang Li	Dalian University of Technology, China
Yuan-Fang Li	Monash University, Australia
Zhenghua Li	Soochow University, China
Hongfei Lin	Dalian University Of Technology, China
Yu Lu	Beijing Normal University, China
Zhunchen Luo	PLA Academy of Military Science, China
Xiaogang Ma	University of Idaho, USA
Yinglong Ma	North China Electric Power University, China
Xian-Ling Mao	Beijing Institute of Technology, China
Peng Peng	Hunan University, China
Guilin Qi	Southeast University, China
Longhua Qian	Soochow University, China
Jing Qiu	Guangzhou University, China
Xipeng Qiu	Fudan University, China
Tong Ruan	East China University of Science and Technology, China
Huawei Shen	Institute of Computing Technology, Chinese Academy of Sciences, China
Wei Shen	Nankai University, China
Chuan Shi	Beijing University of Posts and Telecommunications, China
He Shizhu	Institute of Automation, Chinese Academy of Sciences, China
Dezhao Song	Thomson Reuters, USA
Guojie Song	Peking University, China
Chengjie Sun	Harbin Institute of Technology, China
Weiwei Sun	Peking University, China
Zequn Sun	Nanjing University, China
Buzhou Tang	Harbin Institute of Technology of Shenzhen Graduate School, China
Jie Tang	Tsinghua University, China
Jintao Tang	National University of Defense Technology, China
Siliang Tang	Zhejiang University, China

Fei Tian	Microsoft Research Asia, China
Huaiyu Wan	Beijing Jiaotong University, China
Jing Wan	Beijing University of Chemical Technology, China
Xiaojun Wan	Peking University, China
Jian Wang	Dalian University of Technology, China
Jing Wang	The University of Tokyo, Japan
Li Wang	Taiyuan University of Technology, China
Quan Wang	Institute of Information Engineering, Chinese Academy of Sciences, China
Senzhang Wang	Beihang University, China
Xiao Wang	Tsinghua University, China
Xin Wang	Tianjin University, China
Zhichun Wang	Beijing Normal University, China
Zhigang Wang	Tsinghua University, China
Zhongqing Wang	Soochow University, China
Zhongyu Wei	Fudan University, China
Yunfang Wu	Peking University, China
Xuefeng Xi	Suzhou University of Science and Technology, China
Rui Xia	Nanjing University of Science and Technology, China
Liao Xiangwen	Fuzhou, China
Guohui Xiao	KRDB Research Centre, Free University of Bozen-Bolzano, Italy
Tong Xiao	Northeastern University, China
Yanghua Xiao	Fudan University, China
Xin Xin	Beijing Institute of Technology, China
Bo Xu	Fudan University, China
Hua Xu	AI Speech Co., Ltd., China
Jian Xu	Sun Yat-sen University, China
Jiarong Xu	Zhejiang University, China
Kang Xu	Southeast University, China
Kang Xu	Nanjing University of Posts and Telecommunications, China
Kun Xu	IBM Research, China
Tong Xu	University of Science and Technology of China, China
Yeqiang Xu	Summba Inc., China
Liang Yang	Dalian University of Technology, China
Yang Yang	Zhejiang University, China
Zhihao Yang	Dalian University of Technology, China
Dong Yu	Beijing Language and Culture University, China
Ke Yu	Beijing University of Posts and Telecommunications, China
Ran Yu	L3S Research Center, Germany
Heng Zhang	Huazhong University of Science and Technology, China
Jing Zhang	Renmin University of China, China
Lishan Zhang	Beijing Normal University, China
Songmao Zhang	Chinese Academy of Sciences, China
Weinan Zhang	Harbin Institute of Technology, China
Xi Zhang	Beijing University of Posts and Telecommunications, China

Xiang Zhang	Southeast University, China
Xiaowang Zhang	Tianjin University, China
Xuechao Zhang	National Defence University, China
Zhizheng Zhang	Southeast University, China
Ziqi Zhang	Sheffield University, UK
Sendong Zhao	Harbin Institute of Technology, China
Xin Zhao	Renmin University of China, China
Zhongying Zhao	Shandong University of Science and Technology, China
Zhou Zhao	Zhejiang University, China
Xiaoqing Zheng	Fudan University, China
Chuan Zhou	Institute of Information Engineering, Chinese Academy of Sciences, China
Huiwei Zhou	Dalian University of Technology, China
Junsheng Zhou	Nanjing Normal University, China
Chen Zhu	Baidu Talent Intelligence Center, China
Hou Zhu	Sun Yat-sen University, China
Qili Zhu	Shanghai Jiao Tong University, China
Fuzhen Zhuang	Institute of Computing Technology, Chinese Academy of Sciences, China
Lei Zou	Peking University, China

Sponsors

Platinum Sponsors

LEGAL 华宇元典
INTELLIGENCE 法律智能

美团 美团点评

WeBank 微众银行

Alibaba Group
阿里巴巴集团

Gold Sponsors

Silver Sponsors

Bronze Sponsors

Contents

Towards Answering Geography Questions in Gaokao:
A Hybrid Approach... 1
    Zhiwei Zhang, Lingling Zhang, Hao Zhang, Weizhuo He, Zequn Sun,
    Gong Cheng, Qizhi Liu, Xinyu Dai, and Yuzhong Qu

Distant Supervision for Chinese Temporal Tagging..................... 14
    Hualong Zhang, Liting Liu, Shuzhi Cheng, and Wenxuan Shi

Convolutional Neural Network-Based Question Answering Over
Knowledge Base with Type Constraint 28
    Yongrui Chen, Huiying Li, and Zejian Xu

MMCRD: An Effective Algorithm for Deploying Monitoring
Point on Social Network 40
    Zehao Guo, Zhenyu Wang, and Rui Zhang

Deep Learning for Knowledge-Driven Ontology Stream Prediction 52
    Shumin Deng, Jeff Z. Pan, Jiaoyan Chen, and Huajun Chen

DSKG: A Deep Sequential Model for Knowledge Graph Completion 65
    Lingbing Guo, Qingheng Zhang, Weiyi Ge, Wei Hu, and Yuzhong Qu

Pattern Learning for Chinese Open Information Extraction.............. 78
    Yang Li, Qingliang Miao, Tong Guo, Ji Geng, Changjian Hu,
    and Feiyu Xu

Adversarial Training for Relation Classification with Attention
Based Gate Mechanism 91
    Pengfei Cao, Yubo Chen, Kang Liu, and Jun Zhao

A Novel Approach on Entity Linking for Encyclopedia Infoboxes........ 103
    Xufeng Li, Jianlei Yang, Richong Zhang, and Hongyuan Ma

Predicting Concept-Based Research Trends with Rhetorical Framing 116
    Jifan Yu, Liangming Pan, Juanzi Li, and Xiaoping Du

Knowledge Augmented Inference Network for Natural
Language Inference.. 129
    Shan Jiang, Bohan Li, Chunhua Liu, and Dong Yu

Survey on Schema Induction from Knowledge Graphs. 136
   Qiu Ji, Guilin Qi, Huan Gao, and Tianxing Wu

Author Index 143

Towards Answering Geography Questions in Gaokao: A Hybrid Approach

Zhiwei Zhang, Lingling Zhang, Hao Zhang, Weizhuo He, Zequn Sun, Gong Cheng[✉], Qizhi Liu, Xinyu Dai, and Yuzhong Qu

National Key Laboratory for Novel Software Technology,
Nanjing University, Nanjing, China
{zwzhang,llzhang,haozhang,heweizhuo,zequnsun}@smail.nju.edu.cn,
{gcheng,lqz,daixinyu,yzqu}@nju.edu.cn

Abstract. Answering geography questions in a university's entrance exam (e.g., Gaokao in China) is a new AI challenge. In this paper, we analyze its difficulties in problem understanding and solving, which suggest the necessity of developing novel methods. We present a pipeline approach that mixes information retrieval techniques with knowledge engineering and exhibits an interpretable problem solving process. Our implementation integrates question parsing, semantic matching, and spreading activation over a knowledge graph to generate answers. We report its promising performance on a representative sample of 1,863 questions used in real exams. Our analysis of failures reveals a number of open problems to be addressed in the future.

Keywords: Information retrieval · Knowledge engineering
Natural language processing · Question answering

1 Introduction

Subsequent to the Japanese Todai project [5], China has launched an ambitious initiative to have AI pass the National Higher Education Entrance Examination, commonly known as Gaokao. To meet this grand challenge, advances in AI technologies can in the meantime benefit future-generation computer-aided education, e.g., enabling applications that can interact with learners and act as mentors, or provide assistance to teachers. This long-term goal drives our research of an approach that is capable of not only correctly answering complex questions but also solving problems like human intelligence, i.e., to have an interpretable problem solving process.

Previous research focused on history exams in Gaokao [2,14]. In this paper, we present the first study on geography exams. An example of geography questions is illustrated in Fig. 1, which appears more difficult than history questions for both students and machines due to not only the understanding and exploitation of graphical presentations (e.g., maps and charts) but also the complexity

© Springer Nature Singapore Pte Ltd. 2019
J. Zhao et al. (Eds.): CCKS 2018, CCIS 957, pp. 1–13, 2019.
https://doi.org/10.1007/978-981-13-3146-6_1

and diversity in problem solving. These distinguishing features have turned previous solutions to history questions and existing methods for question answering (QA) ineffective. Specifically, our investigation of more than one thousand real geography questions finds that: quantitative methods (e.g., math calculation) and qualitative methods (e.g., text retrieval) are almost equally important; most questions require mixing them together in a hybrid reasoning process.

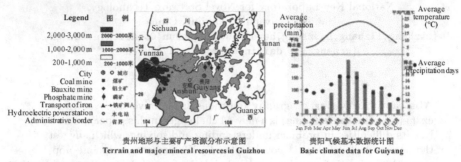

贵州地形与主要矿产资源分布示意图
Terrain and major mineral resources in Guizhou

贵阳气候基本数据统计图
Basic climate data for Guiyang

Questions:

(1) 描述贵州省的地理位置特点。
Q1: Describe the features of the location of Guizhou Province.

(2) 从地形、气候角度说明自然条件对贵州经济发展的不利影响。
Q2: In terms of terrain and climate, explain the negative effects of natural circumstances on the economic development of Guizhou.

(3) 2014年1月，国家级开发区——贵安新区在贵阳和安顺之间正式成立。我国三大电信运营商在此设立云计算中心，中关村科技园、搜狐、百度、京东等企业落户此地，贵安大数据产业基地初具雏形。根据该材料概括贵安大数据产业基地赢得的有利发展条件，结合所学指出该产业基地为吸引企业入驻应进一步采取的措施。
Q3: In January 2014, the Gui-An state-level new area was established, located between Guiyang and Anshun. Over there, the three state-run telecom companies opened their cloud computing centers, and companies including Centek, Sohu, Baidu, and JD set up new branches, showing the initiation of the Gui-An base for the big data industry. With respect to this information, summarize the positive circumstances for developing the Gui-An base for the big data industry, and identify some measures for this industrial base to attract more companies.

Fig. 1. An example of geography question in Gaokao based on a map and a chart.

Therefore, we propose a hybrid approach having a pipeline design to enable interpretable problem solving. Our contributions are threefold.

- We collect, investigate, and make available 1,863 geography questions used in Gaokao or mock exams in recent years. In Sect. 2, we survey the difficulties in answering these questions and summarize the new technical challenges they bring to AI, which can hardly be met by existing methods.
- Our proposed approach mixes information retrieval techniques with knowledge engineering. Section 3 overviews its pipeline. Section 4 describes our current implementation consisting of question parsing, knowledge graph construction, semantic matching, spreading activation and answer generation.
- We carry out extensive experiments on the collected questions, and create a gold standard for the output of each component of the proposed approach. In Sect. 5 we evaluate the approach and its components, show their effectiveness, and analyze their failures.

2   Challenges and Related Work

Differences Between Geography and Other Subjects. Research attention has been given to complex questions appearing in entrance exams at different levels of education including Gaokao. The OntoNova system [1] represents chemistry knowledge in F-Logic, to answer formal queries regarding complex chemical concepts and reactions via rule reasoning. Similar approaches are used to answer physics and biology questions [6]. Different from these natural or formal sciences which require a fully formal method with an emphasis on deep reasoning, history [2] and elementary science questions [3] are solved primarily by using information retrieval techniques supplemented by shallow reasoning.

Geography is distinguished from the above-mentioned subjects in problem understanding and solving. As illustrated in Fig. 1, geography questions are unique because they always involve maps, charts, or other graphical presentations from which information is to be extracted as it is usually essential to problem solving. The understanding of the natural language expressing those questions (e.g., Q1–Q3 in Fig. 1) is distinctively challenged by: the lack of a vocabulary of geographical concepts (e.g., *terrain* and *climate* in Q2), the extensive use of unbounded commonsense knowledge (e.g., *state-level new area* and *big data industry* in Q3), and many new question types that are specific to geography. To solve those questions which often exhibit a hybrid nature, the techniques we need include: (i) all kinds of knowledge representations (e.g., ontological, causal, spatial, and temporal relations) and reasoning capabilities (e.g., logical reasoning, math calculation) which have only been partially considered in answering physics and chemistry questions, (ii) qualitative methods (e.g., text retrieval and summarization) which have only been preliminarily used in answering history questions, and more importantly, (iii) an effective mixture of these methods within a single problem solving process.

Challenges to Traditional Methods for QA. Geography questions in Gaokao are even further beyond the capabilities of traditional methods for QA. None of the above-mentioned challenges can be met by early information retrieval solutions [9] or recent semantic parsing techniques [10] and QA systems [4]. In fact, those methods primarily focus on retrieving answers to short factoid questions, thereby not fitting geography questions in Gaokao. As illustrated by Q3 in Fig. 1, geography questions in Gaokao can be too long to be processed by existing QA solutions, and are often non-factoid expecting answers that are several generated sentences. Other end-to-end methods [12] suffer from lacking interpretability and hence diverge from the long-term goal of our research.

Therefore, not only novel techniques for QA are needed for effective and interpretable solving of geography questions, but also an effort of domain-specific knowledge engineering seems necessary.

3   Overview of the Approach

Towards meeting the identified challenges, we propose a hybrid approach which features a pipeline design and hence an interpretable problem solving process to facilitate future applications. The major design principle is to mix information retrieval techniques (e.g., ranking) with knowledge engineering, considering the sole use of information retrieval is not powerful enough and a fully formal method is unfeasible at the moment due to the complexity in knowledge representation and the difficulty in natural language understanding.

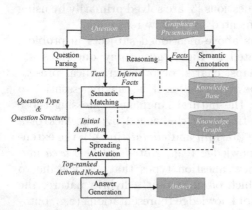

As outlined in Fig. 2, we construct a knowledge graph (KG) for the geography domain. We choose this single but flexible representation as it can help us manage the complexity of knowledge engineering. To answer a question, our approach initially activates a set of nodes in the KG that can find semantic matches in the question text, and then iteratively spreads that activation to neighboring nodes for ranking. Question parsing returns the type and structure of the question, which are used to filter out irrelevant nodes. Finally, the remaining top-ranked nodes are selected, and an answer is generated by concatenating their text.

Fig. 2. Overview of the proposed approach.

We also plan to annotate graphical presentations and produce facts which in turn will be enriched by reasoning over a knowledge base. The resulting facts will also be considered in semantic matching. However, this branch has not been incorporated into our implementation reported in this paper.

4   Detailed Implementation

We collect a representative sample of 1,863 geography questions from Gaokao or mock exams in recent years. The current implementation of our approach is guided by an investigation of this dataset. In this section, we detail our implementation using Q1–Q3 in Fig. 1 as a running example. We also describe the gold standard we create for the output of each component of our approach on that dataset, for characterizing real questions and for evaluation.

The hybrid nature of the approach is highlighted in Sect. 4.4, which is founded on the techniques presented in Sects. 4.1, 4.2 and 4.3.

Table 1. Types of questions

Causes	27.67%	Troubles	4.06%
Circumstances	9.48%	Distribution	4.02%
Effects	8.63%	Changes	3.65%
Features	7.89%	Opinions	1.77%
Comparison	7.71%	Redirection	0.48%
Measures	7.16%	Transport	0.33%
Significance	5.20%	MISC	11.95%

Table 2. Relations in the KG (manual + automated)

State	478+104	Feature	101
Effect	440	Is-a	83+22
Circumstance	365	Factor	27
Measure	257	Manner	16
Aspect	212+103	Action	0+138
		Total	1,979+367

4.1 Question Parsing

Question Decomposition. When applicable, we decompose a question into simpler atomic sub-questions to be separately solved, whose answers will be concatenated in the end. An atomic question in Gaokao is normally identified by an imperative sentence starting with a particular verb. We define 44 such Chinese verbs, e.g., *describe*, *explain*, *summarize*, and *identify*.

Example 1. Q3 is decomposed into two sub-questions.
Q3a: <u>Summarize</u> the positive circumstances for developing the Gui-An base for the big data industry.
Q3b: <u>Identify</u> some measures for this industrial base to attract more companies.

We also use common Chinese coordinating conjunctions (e.g., *and*) and punctuation marks representing coordination to further decompose questions.

Example 2. Q2 is decomposed into two sub-questions.
Q2a: In terms of <u>terrain</u>, explain the negative effects of natural circumstances on the economic development of Guizhou.
Q2b: In terms of <u>climate</u>, explain the negative effects of natural circumstances on the economic development of Guizhou.

Gold Standard 1. *We manually analyze all the 1,863 questions, find 1,155 atomic questions, and decompose the other 708 questions (38%) into 1,556 atomic questions. There are a total of 2,711 atomic questions after decomposition.*

Question Classification and Structuring. To capture the intent of a question, we detect its type and transform it into a type-specific semi-structured tuple representation of its arguments.

We identify 13 question types, plus MISC representing all the others, as shown in Table 1. We detect the type of a question using rules over 84 particular keywords in Chinese. Example rules are as follows.

- A question ending with *circumstances* is a *Circumstances* question.
- A question mentioning *hazards*, *threats*, or *risks* is a *Troubles* question.

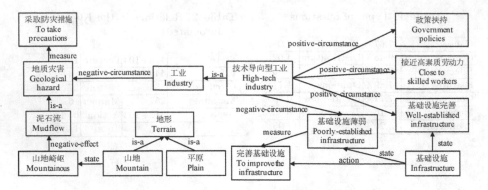

Fig. 3. An example of knowledge graph.

A question will be in MISC if none of the rules are matched.

For each type of question, we define one or several *question structures* to represent the arguments of a question. Example question structures are as follows:

$$\text{Effects (cause, recipient, aspect, polarity)}, \tag{1}$$

$$\text{Circumstances (subject, aspect, polarity)}. \tag{2}$$

The first structure represents a question asking the effects of a particular *polarity* (i.e., positive or negative effects, or unspecified) that the *cause* has on the *recipient*, in terms of some *aspect* which is optional. The second one asks the circumstances which the *subject* is under, of an optional *polarity*, and in terms of an optional *aspect*.

As questions in Gaokao are mainly expressed in a handful of syntactic forms, we transform a question into a question structure based on template matching. An example template is as follows, written as a regular expression.

$$((\text{In terms of})|(\text{Based on}) \ aspect,)?$$
$$\text{explain}|\text{summarize the } polarity? \text{ effects of } cause \text{ on } recipient. \tag{3}$$

Example 3. Q2a is transformed into the question structure in Eq. (1) by Eq. (3):

$$\text{Effects (natural circumstances, the economic development of Guizhou, terrain, negative)}. \tag{4}$$

Q3a is transformed into the question structure in Eq. (2):

$$\text{Circumstances (the Gui-An base for the big data industry, developing, positive)}. \tag{5}$$

Gold Standard 2. *We manually classify all the 2,711 atomic questions. Their distribution is shown in Table 1. We also transform each of the 685 atomic Circumstances, Effects, and Measures questions into a question structure.*

4.2   Knowledge Graph Construction

Our proposed implementation focuses on answering *Circumstances*, *Effects*, and *Measures* questions as they are less dependent on graphical presentations which are not addressed in this paper. These questions constitute 25% of all.

The remaining components of our approach center around a geography knowledge graph (KG) illustrated in Fig. 3, which is semi-structured as each node is a free text that can represent a concept, a fact, or any other thing, whereas each edge represents one of a set of predefined relations. We construct the KG in a semi-automatic way. A domain expert manually constructs the first version to cover *Circumstances*, *Effects*, and *Measures* knowledge in textbooks. The KG is then automatically completed with more edges and nodes to improve its quality.

Manual Construction. The domain expert constructs the KG in four steps. First, geography concepts extracted from textbooks and the official Gaokao guide are added to the KG as nodes. Then, relations extracted from textbooks and other relevant materials are added as edges. These relations, namely *circumstance*, *effect*, and *measure*, provide candidate answers to the corresponding types of questions. Next, more types of relations, e.g., *is-a*, *aspect*, and *state*, are added to connect nodes. Finally, the KG is refined by merging nearly duplicate nodes and substituting some relations (e.g., *effect*) with their sub-relations having polarity information (e.g., *negative-effect*). The resulting KG consists of 1,243 nodes and 1,979 edges. Table 2 shows its distribution of relations.

Gold Standard 3. *We manually map the correct answer of each* Circumstances, Effects, *and* Measures *question to a set of nodes in the KG, for evaluating the output of our approach.*

Automated Completion. To complement human efforts, we devise an algorithm for automatically completing the KG. Concepts mentioned in existing nodes are recognized. If a concept has no corresponding node in the KG, it will be added as a new node. The concept is connected to the nodes mentioning it via proper relations.

Specifically, concepts are recognized using predefined *POS patterns* tagged by the Stanford Parser. For example, given a node text (in Chinese) tagged with a sequence of NNs followed by VA, these NNs will identify a (new) concept node, which will be connected to the original node via *state*. In this way, 241 nodes (+19%) and 367 edges (+19%) are automatically added to the KG, as detailed in Table 2.

Example 4. From the node *Poorly-established infrastructure* which (in Chinese) is tagged with NN NN VA, *Infrastructure* (tagged with NN NN) is recognized as a (new) node, and is connected to *Poorly-established infrastructure* via *state*.

4.3   Semantic Matching

We initially activate all the nodes in the KG that are matched with the question text. A node will be activated if its content words/phrases are all matched by the question text. We realize semantic matching by expanding each content word/phrase with its synonyms and hyponyms. Synonyms are imported from an online Chinese thesaurus. Hyponyms are obtained from a hierarchy of 960 domain terms defined by a domain expert.

Example 5. The node *Terrain* in Fig. 3 is matched with the word *terrain* in Q2a. The node *High-tech industry* in Fig. 3 is matched with its hyponym *big data industry* in Q3a.

Gold Standard 4. *We manually match the text of all the* Circumstances, Effects, *and* Measures *questions with nodes in the KG, and label a total of 3,823 matches.*

4.4   Spreading Activation and Answer Generation

We concatenate the text of top-ranked nodes in the KG as an answer. Nodes are ranked by closeness to the initial activation found by semantic matching. For ranking, we use a spreading activation method (SA) called random walk with restart (RWR) [13]. SA iteratively spreads the amount of activation received by each node to its neighbors until convergence. During spreading, RWR allows to jump back to an initially activated node, thereby lifting the ranking of the nodes that are closer to the initial activation. This is the key step of our approach as it mixes information retrieval techniques (i.e., SA and RWR ranking) with knowledge engineering (i.e., KG) to discover implicit answers.

Example 6. As Q2a matches the node *Terrain* in Fig. 3, nodes that are closer to it will be ranked higher, e.g., *Geological hazard* compared with *Poorly-established infrastructure.*

Further, we leverage the structure of a question to filter out irrelevant nodes. For each structure we define one or several *relation path expressions* for traversing the KG, which are written as regular expressions. For example, we define the following relation path expression for the question structure in Eq. (2) when *polarity* is positive:

$$subject \text{ is-a}^* \text{ positive-circumstance}, \tag{6}$$

which starts searching the KG from each node matched by the *subject* argument, passes through any number of *is-a* edges, and finally passes through a *positive-circumstance* edge and reaches a set of nodes which will be labeled as relevant; if *subject* does not match any node, search will start from every node. Irrelevant nodes will be removed from the answer. This step realizes answer type coercion with reasoning capabilities, e.g., is-a reasoning in this example.

Table 3. Percentage of correctly classified questions

Causes	99.33%	Troubles	97.27%
Circumstances	98.83%	Distribution	97.25%
Effects	97.44%	Changes	100.00%
Features	93.46%	Opinions	77.08%
Comparison	93.78%	Redirection	92.31%
Measures	91.75%	Transport	77.78%
Significance	99.29%	MISC	86.42%
		Overall	95.50%

Table 4. Percentage of correctly structured questions

Circumstances	88.72%
Effects	89.32%
Measures	75.26%
Overall	85.11%

Example 7. As the *subject* argument of Q3a's question structure in Eq. (5) matches the node *High-tech industry* in Fig. 3, using the relation path expression in Eq. (6) we identify three relevant nodes: *Government policies, Close to skilled workers*, and *Well-established infrastructure*.

The proposed problem solving process is interpretable. We can extract from the KG those paths connecting returned nodes with initially activated nodes (which are matched with the question text), and present as answer explanations.

5 Evaluation

We evaluated our approach based on the collected questions and the gold standard we created. All the relevant resources have been made available at: http://ws.nju.edu.cn/gaokao/ccks-18.zip.

5.1 Evaluation of the Entire Approach

Evaluation Design. In Gold Standard 3 we labeled a set of nodes in the KG as the correct answer to a question. Based on that, we configured both our approach and a baseline method to output k top-ranked nodes for each of the 685 questions, and calculated their average F1 scores on all the questions.

Baseline. We compared with BM25, a baseline widely used in information retrieval. To fit our problem, we adapted it to take nodes from *circumstance, effect*, and *measure* relations in the KG as candidate answers to *Circumstances, Effects*, and *Measures* questions, respectively, and rank nodes by the BM25 relevance between node and question text. On the other hand, semantic parsing and other knowledge-based methods for QA were not compared because they could not process our questions due to the unfitness discussed in Sect. 2.

Results. As shown in Fig. 4, the F1 score of the baseline was consistently below 0.1 under any setting of k. The overall F1 score of our approach reached 0.26 when $k = 5$, and exceeded 0.3 on *Circumstances* and *Measures* questions, significantly outperforming the baseline.

5.2   Evaluation of Components

Although largely exceeding the baseline, there was considerable room for improving our approach. We evaluated each component to identify their limitations.

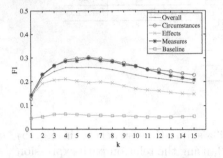

Fig. 4. F1 score of k top-ranked nodes as an answer.

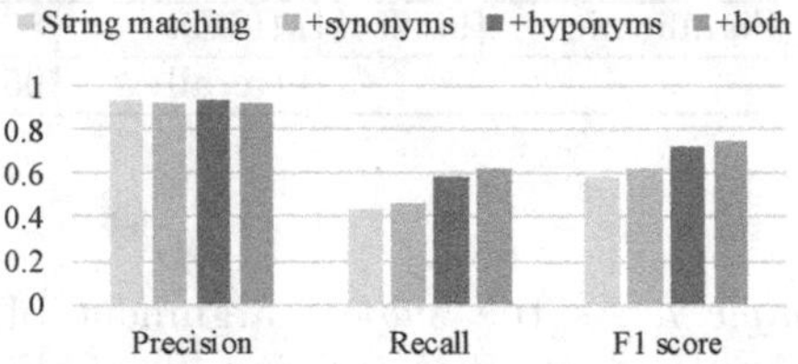

Fig. 5. Precision, recall, and F1 score of matching.

Question Decomposition. Among the 1,155 indecomposable atomic questions in Gold Standard 1, our method mistakenly decomposed 40 (3%) as they contained triggering verbs or coordinating conjunctions. Among the other 708 non-atomic questions, the decomposition outputted by our method was exactly the same as the gold standard on 595 (84%); the generated sub-questions contained insignificant errors on 105 (15%), mainly due to incorrect phrase structure trees about coordinating conjunctions; the decomposition contained significant errors only on 8 (1%). To conclude, out of the 1,863 questions, the automated decomposition of 1,815 (97%) was (nearly) correct.

Question Classification and Structuring. Among the 2,711 atomic questions in Gold Standard 2, our method correctly classified 2,589 (96%). As shown in Table 3, the accuracy exceeded 90% on most question types, though relatively low on *Opinions* and *Transport* questions because they were given low priority when a question could match multiple rules leading to different types.

Among the 685 *Circumstances*, *Effects*, and *Measures* questions in Gold Standard 2, our method transformed 583 (85%) into correct question structures, as shown in Table 4. Incorrect transformation was mainly due to incorrect word segmentation, an open problem for Chinese processing.

Semantic Matching. In Gold Standard 4 we labeled 3,823 matches. Based on that, we calculated the average precision, recall, and F1 score of the matches generated by our method on all the questions. As shown in Fig. 5, the precision of string matching was high (0.94), but the recall was unsatisfactory (0.43). Expanding content words/phrases with synonyms and hyponyms, recall notably increased by 0.03 and 0.16, respectively, whereas the decreases in precision were negligible, so F1 scores rose. Using both synonyms and hyponyms, F1 score reached 0.75. Our method mainly failed to find partial matches and deep semantic matches. We did not allow partial matches because otherwise precision would drop largely. To identify deep semantic matches representing implicit relatedness, e.g., the match between the text *state-level new area* and the node *Government policies* in the KG, more sophisticated semantic matching techniques would be needed, or the KG should be substantially extended to cover more concepts and relations such that implicit matches would more likely be identified during spreading activation.

Spreading Activation and Answer Generation. In addition to the evaluation of the entire approach in Sect. 5.1, we also evaluated the k top-ranked nodes outputted by our spreading activation method using manually labeled matches in Gold Standard 4 as initial activation. F1 score was raised to 0.31 (from 0.26) when $k = 5$, still not very high mainly due to the incompleteness of the KG. Many concepts mentioned in question text were not covered by the KG, thereby incapable of effectively guiding the spreading activation.

Concluding Remarks. Our question parsing, though based on a few simple rules and templates, generated high-quality results. Semantic matching was satisfactory but could be improved. The incompleteness of the KG became a bottleneck of our current implementation.

6 Future Directions

We present the first research effort to answer geography questions in Gaokao. We have identified the difficulties in problem understanding and solving, and we invite the community to continue contributing solutions to this new AI challenge, as it will benefit future-generation computer-aided education and push the boundaries of AI. To provide supporting resources, we have made all our data publicly available.

Our major technical contribution is a pipeline approach which mixes information retrieval techniques with knowledge engineering and realizes interpretable problem solving. Effectively integrating question parsing, semantic matching, spreading activation and answer generation over a constructed knowledge graph, our approach exhibits a non-trivial capacity to answer complex questions and significantly outperforms a baseline in an extensive experiment. The result is promising as graphical presentations in questions and more powerful reasoning capabilities are still yet to be exploited.

Several open problems have surfaced to be addressed in future work. (i) To extend the coverage of the KG, we are seeking practically effective methods for automated construction of KG. Recent studies have shown it is possible to automatically extract specific relations of reasonably good quality from text to answer the corresponding type of questions, e.g., causality [7]. We will adapt those methods to build a large-scale KG automatically and evaluate their effectiveness. (ii) Although KG as a single flexible representation has successfully helped us manage the complexity of knowledge engineering, it is unlikely to fit all question types. We will consider and integrate other knowledge representations which are possibly associated with different types of reasoning. (iii) Despite the effectiveness of rule- and template-based question parsing in our application, more powerful and robust techniques are still needed to process user-generated questions in potential future applications which are less formal than those in Gaokao. In particular, to better understand and use more information (e.g., newly defined concepts and laws) provided in question text, we will experiment with learning [8] and semantic parsing methods [11] which have been used to solve verbally expressed number problems.

Acknowledgements. This work was supported in part by the NSFC under Grants 61772264 and 61572247, in part by the 863 Program under Grant 2015AA015406, and in part by the Qing Lan and Six Talent Peaks Programs of Jiangsu Province.

References

1. Angele, J., Moench, E., Oppermann, H., Staab, S., Wenke, D.: Ontology-based query and answering in chemistry: OntoNova project Halo. In: Fensel, D., Sycara, K., Mylopoulos, J. (eds.) ISWC 2003. LNCS, vol. 2870, pp. 913–928. Springer, Heidelberg (2003). https://doi.org/10.1007/978-3-540-39718-2_58
2. Cheng, G., Zhu, W., Wang, Z., Chen, J., Qu, Y.: Taking up the Gaokao challenge: an information retrieval approach. In: Proceedings of the Twenty-Fifth International Joint Conference on Artificial Intelligence, pp. 2479–2485. International Joint Conferences on Artificial Intelligence, New York, July 2016
3. Clark, P., Etzioni, O., Khot, T., Sabharwal, A., Tafjord, O., Turney, P.: Combining retrieval, statistics, and inference to answer elementary science questions. In: Proceedings of the Thirtieth AAAI Conference on Artificial Intelligence, pp. 2580–2586. AAAI Press, Phoenix, February 2016
4. Ferrucci, D., et al.: Building Watson: an overview of the DeepQA project. AI Mag. **31**(3), 59–79 (2010)
5. Fujita, A., Kameda, A., Kawazoe, A., Miyao, Y.: Overview of Todai robot project and evaluation framework of its NLP-based problem solving. In: Proceedings of the Ninth International Conference on Language Resources and Evaluation, pp. 2590–2597. European Language Resources Association, Reykjavik, May 2014
6. Gunning, D., et al.: Project Halo update - progress toward digital aristotle. AI Mag. **31**(3), 33–58 (2010)
7. Hidey, C., McKeown, K.: Identifying causal relations using parallel Wikipedia articles. In: Proceedings of the Fifty-Fourth Annual Meeting of the Association for Computational Linguistics, pp. 1424–1433. Association for Computational Linguistics, Berlin, August 2016
8. Hosseini, M.J., Hajishirzi, H., Etzioni, O., Kushman, N.: Learning to solve arithmetic word problems with verb categorization. In: Proceedings of the 2014 Conference on Empirical Methods in Natural Language Processing, pp. 523–533. Association for Computational Linguistics, Doha, October 2014
9. Kolomiyets, O., Moens, M.F.: A survey on question answering technology from an information retrieval perspective. Inf. Sci. **181**(24), 5412–5434 (2011)
10. Liang, P.: Learning executable semantic parsers for natural language understanding. Commun. ACM **49**(9), 68–76 (2016)

11. Shi, S., Wang, Y., Lin, C.Y., Liu, X., Rui, Y.: Automatically solving number word problems by semantic parsing and reasoning. In: Proceedings of the 2015 Conference on Empirical Methods in Natural Language Processing, pp. 1132–1142. Association for Computational Linguistics, Lisbon, September 2015

12. Tan, M., dos Santos, C., Xiang, B., Zhou, B.: Improved representation learning for question answer matching. In: Proceedings of the Fifty-Fourth Annual Meeting of the Association for Computational Linguistics, pp. 464–473. Association for Computational Linguistics, Berlin, August 2016

13. Tong, H., Faloutsos, C.: Center-piece subgraphs: problem definition and fast solutions. In: Proceedings of the Twelfth ACM SIGKDD International Conference on Knowledge Discovery and Data Mining, pp. 404–413. Association for Computing Machinery, Philadelphia, August 2006

14. Yu, K., Liu, Q., Zheng, Y., Zhao, T., Zheng, D.: History question classification and representation for Chinese Gaokao. In: Proceedings of the Twentieth International Conference on Asian Language Processing, pp. 129–132. Institute of Electrical and Electronics Engineers, Tainan, November 2016

Distant Supervision for Chinese Temporal Tagging

Hualong Zhang, Liting Liu, Shuzhi Cheng, and Wenxuan Shi[✉]

Nankai University, Tianjin, China
nankaizhl@gmail.com,
nkliuliting826@mail.nankai.edu.cn,
shuzhichengspace@163.com, shiwx@nankai.edu.cn

Abstract. Temporal tagging plays an important role in many tasks such as event extraction and reasoning. Extracting Chinese temporal expressions is challenging because of the diversity of time phrases in Chinese. Usually researchers use rule-based methods or learning-based methods to extract temporal expressions. Rule-based methods can often achieve good results in certain types of text such as news but multi-type text with complex time phrases. Learning-based methods often require large amounts of annotated corpora which are hard to get, and the training data is difficult to extend to other tasks with different text type. In this paper, we consider time expression extraction as a sequence labeling problem and try to solve it by a popular model BiLSTM+CRF. We propose a distant supervision method using CN-DBPedia (an open domain Chinese knowledge graph) and BaiduBaike (one of the largest Chinese encyclopedias) to generate a dataset for model training. Results of our experiments on encyclopedia text and TempEval2 dataset indicate that the method is feasible. While obtaining acceptable tagging performance, our approach does not involve designing manual patterns as rule-based ones do, does not involve the constructing annotated data manually, and has a good adaptation to different types of text.

Keywords: Chinese temporal tagging · Distant supervision · Knowledge graph

1 Introduction

Temporal information always needs to be extracted from text for its importance in many NLP tasks such as event extraction and reasoning. So far, research on extraction and normalization of temporal expressions mainly focus on western languages like English, French and Spanish. Chinese temporal tagging especially for some long time expressions is challenging and less explored. Chinese has different time systems, including: Chinese lunar system, TianGan-DiZhi (GZ) system, Jie-Qi and Regnal year system [1]. Unlike simple digital temporal expressions, each time system in Chinese has its unique vocabulary. Some are related to the ancient dynasty and names of emperors, for example "丙申猴年，清光绪二十二年" (Bingshen year, the year of monkey, the 22nd year of Emperor Guangxu in the Qing Dynasty). And some are related to historical events for example "建国初期" (Early founding of the People's Republic of China). Figure 1 shows that the time phrase "1896年" can be easily

© Springer Nature Singapore Pte Ltd. 2019
J. Zhao et al. (Eds.): CCKS 2018, CCIS 957, pp. 14–27, 2019.
https://doi.org/10.1007/978-981-13-3146-6_2

marked out by general temporal taggers, for example the HeidelTime online demo, while the whole temporal expression in the sentence is "1896年农历丙申年 (猴年) 清光绪二十二年".

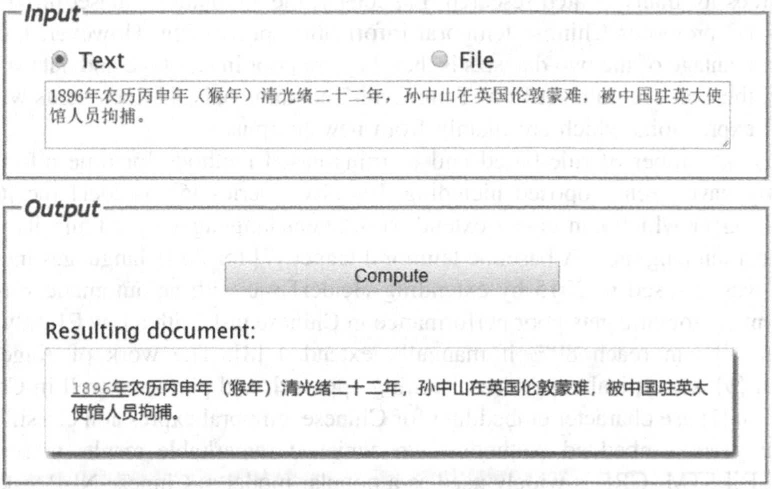

Input

◉ Text   ◯ File

> 1896年农历丙申年（猴年）清光绪二十二年，孙中山在英国伦敦蒙难，被中国驻英大使馆人员拘捕。

Output

Compute

Resulting document:

> 1896年农历丙申年（猴年）清光绪二十二年，孙中山在英国伦敦蒙难，被中国驻英大使馆人员拘捕。

Fig. 1. Example of Chinese temporal tagging with HeidelTime online system.

In this paper, we only focus on temporal expression labeling without classification and normalization. Usually researchers use rule-based methods or learning-based methods to extract temporal expressions from text. Rule-based methods can often perform well in the certain corpus by well-designed rules. Learning-based methods often require expensive annotated data for training. And the model trained with certain types of text is difficult to extend to other types of text, because it's hard for us to annotate all types of possible contexts and temporal phrases. In this paper, we design a distant supervision method with Chinese knowledge graph (CN-DBPedia) [2]. We use the time-related property-value pairs in the Chinese knowledge graph to generate annotation data. Word segmentation and POS tagging are used to make up for the performance loss caused by de-noising. Then a popular sequence labeling model BiLSTM+CRF is used to complete temporal tagging. The corpus for our model training comes from BaiduBaike (one of the largest Chinese encyclopedias). We test the model on text from the same source, and the performance comparable to supervised methods indicates that our method is feasible. Testing on the TempEval2 dataset, the F1 score of our tagging still reaches 72.13%. This indicates that our model has good extensibility for different types of text.

2 Related Works

There are two well-known evaluation tasks involving Chinese time expression extraction, namely, ACE2005 [3] and TempEval-2 [4], which are also the two main data sources for many related research. Especially, the annotation dataset provided by TempEval-2 promotes Chinese temporal information processing. However, the common disadvantage of the two datasets is that they are poor in text type and data size. For example, the Chinese train set from TempEval-2 contains only 44 documents with 746 temporal expressions which are mainly from news corpus.

So far, a number of rule-based and learning-based methods for time information processing have been proposed including TempEval series [5]. HeidelTime [6] is a temporal tagger which can easily extend to different languages by editing its pattern files for certain language. A baseline temporal tagger [7] for 200+ languages including Chinese was released in 2015 by extending HeidelTime with an automatic way. The HeidelTime-automatic gets poor performance in Chinese task with a low F1 only 7.6%, while the F1 can reach 89% if manually extended [8]. The work of Angeli and Uszkoreit [9] is a typical supervised learning approach and performs well in Chinese. Yin and Jin [1] use character embedding for Chinese temporal expression classification. In recent years, embedded methods have achieved remarkable results in sequence labeling. BiLSTM+CRF is widely used as a popular model in Chinese NLP tasks such as entity recognition [10] and sentiment analysis [11]. In this paper, we will use character-based BiLSTM+CRF for temporal labeling. With the development of large knowledge bases like Freebase [12], Yago [13] and DBPedia [14], distant supervision methods are increasingly used in event extraction, relation recognition and other tasks. In 2009, Mintz et al. put forward a distant supervision relation extraction method based on Freebase [15]. In 2016, Huang et al. studied the de-noising of distant supervision methods in Chinese relation extraction [16].

3 Methodology

In this section, we will introduce our temporal tagging approach in three parts:

- **Task Description.** We make clear the target and rules of the temporal tagging task.
- **Train Set Construction.** CN-DBPedia triples and POS information will be used to annotate the sentences from BaiduBaike to be our train set for sequence labeling model.
- **Sequence Labeling Model.** A sentence is encoded by a bidirectional Long Short-term Memory Network (BiLSTM) and the label for each character in the sentence will be predicted by a Conditional Random Field (CRF) layer. We train this BiLSTM+CRF model with self-built data annotated in our distant supervision way.

3.1 Task Description

Temporal tagging in this paper means that given a Chinese sentence S, we should mark out all the time expressions $\{t_1, t_2, \cdots, t_n\}$ contained in the sentence. Without word segmentation, here we use IOB (Inside, Outside, Beginning) tagging schema to label each character in the sentence, as shown in Fig. 2.

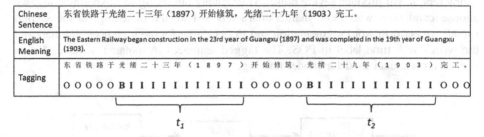

Fig. 2. Example of temporal tagging for a Chinese sentence with IOB tagging schema.

For temporal expression, we follow TIMEX3 tags in TimeML [17], and use typical types including Date, Time, Duration and Set (set of times). That is to say all the Chinese phrases that describe a time point or time range should be considered, listed as Table 1. For long expressions like the one in Fig. 1, we tend to consider it as a whole without splitting.

Table 1. Examples of typical Chinese temporal expressions.

Type	Example
Date	2017年9月11日 (September 11, 2017) 今天 (today) 农历正月十五 (the 15th day of the first month of the Chinese lunar calendar)
Time	凌晨3点半 (3:30 in the morning) 日落时分 (at sunset) 第五天上午 (the fifth day morning)
Duration	昨天一整天 (all day yesterday) 第二次世界大战期间 (during the Second World War) 上千年 (thousands of years)
Set	每周五 (every Friday) 节假日 (on holidays) 每年端午节 (at the Dragon Boat Festival every year)

3.2   Train Set Construction

Dataflow and Architecture

Figure 3 shows the dataflow and architecture of our train set constructor. Firstly we need to collect a large number of multi-style sentences from BaiduBaike. In addition, we pick out all the time-related triples from CN-DBPedia and construct the Time Expression Candidate Set in Fig. 3 with the property values of these triples. Then batch scripts take two rounds of annotation for the sentences. In the first round, we mark out long temporal expressions in the sentences by simple text matching. In practice, if the length of the expressions we match are too long, we will get very few cases for training. If we directly use text matching to find expressions that contain only one or two characters, it will produce severe noise. After simply observing and experimenting, we choose candidates with more than 5 characters to take the matching. In the second round, we perform word segmentation and POS tagging for the sentences and mark out the words with time label in POS. The tagged sentences, Annotated Sentence Set in Fig. 3, are the data used for model training.

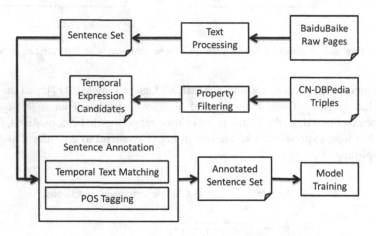

Fig. 3. Dataflow and architecture of train set construction.

The reason we use two rounds of annotation is that we find simply matching short temporal candidates in sentences will cause a lot of mistakes, because there are serious ambiguities in short expressions in Chinese. For example, "宋" can be a temporal word representing the Song Dynasty while it can also be the family name of a person. As for the time information provided by POS tagging, we find that it is good at discovering time words in context while useless for long expressions. This feature is a powerful complement to our annotation work.

Text Processing
Existing manually annotated temporal corpora have problems with insufficient types and limited size. Take the Chinese training set from TempEval-2 as an example, it contains only 44 documents with 746 temporal expressions and the sentences are mainly from news which is hard to adapt to varied contexts in Chinese.

As can be seen from Fig. 3, the texts we use come from BaiduBaike. BaiduBaike is one of the largest Chinese encyclopedia websites with a sufficient number of all types of Chinese sentences in semi-structured data and unstructured text. After removing html and other script elements, more than 200,000 lines are extracted from 30,000 randomly selected BaiduBaike pages. We only collect sentences in Chinese, in other words, the western characters in a sentence must less than 50%. Besides, sentences less than 10 characters are not included. After filtering, more than 100,000 sentences are used for the experiment.

Property Filtering

It is difficult for us to enumerate a large number of time expression patterns manually. CN-DBPedia is an open domain Chinese knowledge graph which contains a large number of time related properties of entities. These entities and properties are stored in triples, the format is <Entity_1, Property/Relation, Value/Entity_2>, for example <李小龙 (Bruce Lee), 出生日期 (Birth Date), 1940年11月27日 (November 27, 1940)>. We screen out triplets with base time words in their second elements (Property/Relation) as temporal triples. The third element of a temporal triple is regarded as a temporal expression as shown in Table 2. From the candidates with more than 5 characters, we keep the 200,000 most frequently occurring temporal expressions.

Table 2. Base time words for property filtering.

Base time words	Triple example	Temporal expression in example
时间(time)	<国际禁毒日, 起源时间, 20世纪80年代>	20世纪80年代
日期(date)	<"恶鹰II", 处理日期, 2004.02.18>	2004.02.18
年代/时代 (age)	<龙口古墓群, 所属年代, 西晋至唐代>	西晋至唐代
年份(year)	<静阳村太清观, 创建年份, 元至大年 (1038年)>	元至大年(1038年)
月份(month)	<鬼宿, 对应月份, 阳历6月22日—7月7日>	阳历6月22日—7月7日
朝代 (dynasty)	<元朝经济, 朝代时间, 1271年—1368年>	1271年—1368年
时长 (duration)	<#thatPower, 歌曲时长, 4分39秒>	4分39秒
周期(period)	<采购排行榜, 时间周期, 最近30天>	最近30天

Sentence Annotation

The final step in constructing training data is to annotate the collected sentences as shown in Fig. 3. We take two rounds of annotation for the sentences to ensure a high accuracy in both long temporal expressions and short ones.

Temporal Text Matching

We collect a large number of time expressions from CN_DBPedia, both in digital and complex Chinese forms. We use long expressions which are more than five characters for text matching to reduce errors caused by phrase ambiguity or improper segmentation. The process of this annotation round can be learned from the example in Fig. 4.

The digits in sentences will cause the problem of too large search space. So before matching the text, we replace the digits with '#' both in sentences and time expressions. We take expressions from the set to match a sentence with a length-first strategy. The

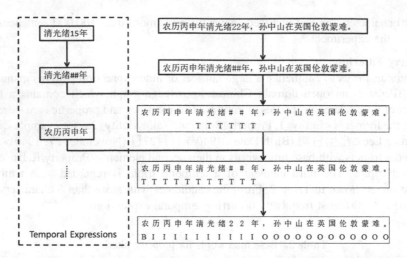

Fig. 4. The process of text match with temporal expressions from CN-DBPedia triples.

characters of a match in the sentence are marked with 'T'. Finally, the complete matches in the sentence will be annotated in the form of IOB like the example in Fig. 4.

POS Tagging

Chinese POS tagging can efficiently find out some commonly used time words like "今天" (today), "古代" (ancient) and "现在" (now). The two temporal expressions in Fig. 5 Example 1 are both marked out with 't'. Although we find it to be invalid for long time expressions shown as Example 2 in Fig. 5, it is a strong supplement to the results of our first round annotation. In practice, we use jieba (a popular Chinese NLP tool) to perform POS tagging for the sentences. Then, the words marked with 't' are added to the original data as newly discovered time expressions.

Example 1

古代	咸阳	地理位置	在	渭河	两岸	，	涵盖	今天	西安	和	咸阳	部分	区域	。
t	ns	n	p	ns	f	x	v	t	ns	c	ns	n	n	x

Example 2

咸阳	是	秦孝公	十二年	（	公元前	350	年）	商鞅	变法	后	秦国	的	都城	。
ns	v	nr	m	x	t	m	m x	nr	vn	f	ns	uj	ns	x

Fig. 5. Examples of Chinese POS tagging result with jieba.

3.3 Sequence Labeling Model

Huang et al. explored the performance of different LSTM-based models in sequence labeling in 2015 [18]. BiLSTM+CRF has excellent performance in tasks such as entity recognition and POS tagging. Further, the use of character-based BiLSTM+CRF also achieves good results in Chinese tasks [10]. In this paper, to complete the temporal tagging we use BiLSTM+CRF similar to the architecture proposed in [19] to be our labeling model. In this section we will briefly introduce BiLSTM, CRF and their integration.

BiLSTM

Recurrent neural networks (RNNs) are a family of neural networks designed for sequential data processing. RNNs take a sequence of vectors $(x_1, x_2, ..., x_n)$ as input and return another sequence $(h_1, h_2, ..., h_n)$ that represents some state information about the sequence at each step of the input. Due to the vanishing gradient problem, general RNNs can't preserve distant historical information well [20]. LSTMs incorporate a memory-cell to combat this issue and have shown great capabilities to capture long-range dependencies. An LSTM cell consists of an input gate, an output gate and a forget gate. Both input gate and forget gate results are used to update the cell state. Formulated as follows:

$$
\begin{aligned}
i_t &= \sigma(W_{xi}x_t + W_{hi}h_{t-1} + W_{ci}c_{t-1} + b_i) & \text{(input gate)} \\
f_t &= \sigma(W_{xf}x_t + W_{hf}h_{t-1} + W_{cf}c_{t-1} + b_f) & \text{(forget gate)} \\
c_t &= f_t \odot c_{t-1} + i_t \odot \tanh(W_{xc}x_t + W_{hc}h_{t-1} + b_c) & \text{(cell state)} \\
o_t &= \sigma(W_{xo}x_t + W_{ho}h_{t-1} + W_{co}c_t + b_o) & \text{(output gate)} \\
h_t &= o_t \odot \tanh(c_t) & \text{(output)}
\end{aligned}
$$

where σ is the element-wise sigmoid function, $\odot$ is the element-wise product, W's are weight matrices, and b's are biases.

A bidirectional LSTM (BiLSTM) is used to encode the context vector of a character in the sentence. Given a sentence S containing n characters $(x_1, x_2, ..., x_n)$, each character represented as a d-dimensional vector, an LSTM computes a representation $\overrightarrow{h_t}$ of the left context of the sentence at every character t. In the same way, by reading the same sentence in reverse, we can get another LSTM which computes the right context $\overleftarrow{h_t}$ starting from the end of the sentence. The former is called forward LSTM and the latter is called backward LSTM. The context vector of a character is obtained by concatenating its left and right context representations, $h_t = \left[\overrightarrow{h_t}, \overleftarrow{h_t}\right]$.

CRF

We use CRF which utilizes features found in a certain context window to model the outputs of the whole sentence. For an input sentence $X = (x_1, x_2, ..., x_n)$, P is the matrix of scores outputted by BLSTM network size $n \times k$, where k is the number of distinct tags (I, B and O). $P_{i,j}$ is the score of the jth tag of the ith character in a sentence.

For a sequence of predictions $y = (y_1, y_2, ..., y_n)$, we define its score to be

$$
s(X, y) = \sum_{i=0}^{n} A_{y_i, y_{(i+1)}} + \sum_{i=1}^{n} P_{i, y_i}
$$

where A is a matrix of transition scores such that $A_{i,j}$ represents the score of a transition from the tag i to tag j. y_0 and y_n are added to the set of possible tags as the start and end tags of a sentence. The transition matrix A is a square matrix of size $k + 2$. A softmax layer is applied over all possible tag sequences, and the probability for a sequence y is

$$p(y|X) = \frac{e^{s(X,y)}}{\sum_{\tilde{y} \in Y_X} e^{s(X,\tilde{y})}}.$$

When training the model, we maximize the log-probability of the correct tag sequence:

$$\log(p(y|X)) = s(X,y) - \log(\sum_{\tilde{y} \in Y_X} e^{s(X,\tilde{y})})$$

$$= s(X,y) - \log \underset{\tilde{y} \in Y_X}{add}\, s(X,\tilde{y})$$

where Y_X represents all possible tag sequences for a sentence X including those that are not in IOB format. From the formulation above, it is evident that invalid sequences of output labels will be discouraged. When decoding, we predict the output sequence that obtains the maximum score given by:

$$y* = \arg\max_{\tilde{y} \in Y_X} s(X,\tilde{y})$$

The formula above can be computed with dynamic programming when decoding, since we only modeling bigram interactions between outputs.

The architecture of the whole sequence labeling model is shown in Fig. 6. The characters in the sentence are previously converted to vectors before we feed them to the BiLSTM layer. The vector representing each character is generated in advance by a

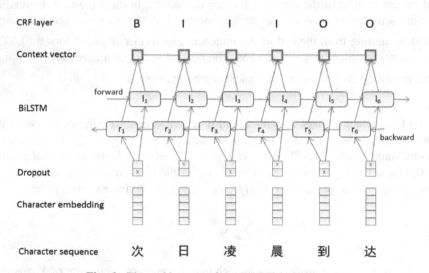

Fig. 6. The architecture of the BiLSTM+CRF model

word2vec model trained with wiki Chinese corpus. For more details of the implementation, we will introduce in the experimental section.

4 Experiments and Analysis

4.1 Model Settings

Since we use BiLSTM+CRF as our sequential labeling model, some basic settings are involved in training and prediction such as batch size, learning rate, and dropout rate. The focus of our research is not to optimize the network structure, for different data groups we will use the same typical model settings listed in Table 3. In practice the training process will stop if there is no improvement for five consecutive epochs though we set 50 as the max train epoch.

Table 3. Model settings when using BiLSTM+CRF.

Model setting	Value
Input character vector dimension	100
Batch size	64
Number of BiLSTM layer	1
Number of nodes	256
Dropout rate	0.5
Loss function	Loss = Log likelihood+L2
Size of validation set	10%
Learning rate	0.001
Max train epoch	50

4.2 Experiment Grouping

We build a train set using our distant supervision approach. To verify the feasibility of our method, we manually tagged a test set using sentences from the same source, BaiduBaike open text. In addition, we used TempEval-2's Chinese data set as a comparison to observe the scalability of our temporal tagger. The role played by the POS tagging will also be explained in experiments. Table 4 lists the data details for each group of experiments.

4.3 Results and Analysis

We use script conlleval.pl provided by CRF++ to score the results of each group and mainly focus on P, R and F1 measures listed in Fig. 5. Because temporal expressions often do not occupy too much in a sentence and most of the character marks are O, the accuracy of the tagging is high but meaningless.

Table 4. Data details for each group of experiments.

Group name	Train set		Test set	
	Content	Size (sentence)	Content	Temporal expressions
TE_2_TE	TempEval2 train set	932	TempEval2 test set	190
TE_2_open	TempEval2 train set	932	Our test set	894
CTTKG_2_TE_no_POS	Our train set no POS tagging	Over 100,000	TempEval2 all data	936
POS_2_TE	Our train set only POS tagging	Over 35000	TempEval2 all data	936
POS_2_open	Our train set only POS tagging	Over 35000	Our test set	894
CTTKG_2_open	Our train set	Over 100,000	Our test set	894
CTTKG_2_TE	Our train set	Over 100,000	TempEval2 all data	936

The results in Table 5 confirm the viewpoint of this paper from several aspects. The first group TE_2_TE indicates that the BiLSTM+CRF model we choose performs well when we have high quality training data. However when we test the model trained by TempEval-2 with test data from another corpus, the F1 score is extremely low. This is also the problem we want to solve, that is, the data type of manual annotation is limited which makes it difficult to extend to different corpora. Our train set constructed with open domain knowledge graph is a lot better. Testing on the homologous corpus, the P, R and F1 of our method all exceed 75% as group CTTKG_2_open in Table 5 shows. Comparing CTTKG_2_TE with TE_2_open, POS_2_TE and CTTKG_2_TE_no_POS, our two-round-annotation strategy has obvious advantages in generalization performance.

Table 5. Statistical evaluation for each group of experimental results.

Group name	B			I			Total			Accuracy
	P	R	F1	P	R	F1	P	R	F1	
TE_2_TE	81.75%	79.84%	80.78%	88.89%	83.72%	86.23%	86.45%	82.43%	84.39%	98.75%
TE_2_open	18.74%	23.33%	20.79%	93.91%	39.99%	56.09%	70.49%	37.76%	49.17%	88.46%
CTTKG_2_TE_no_POS	56.17%	19.69%	29.17%	68.40%	56.03%	61.60%	66.50%	45.00%	53.68%	95.60%
POS_TE	77.87%	43.53%	55.84%	79.16%	23.09%	35.75%	78.57%	29.29%	42.68%	95.34%
POS_2_open	29.52%	5.55%	9.34%	42.13%	1.31%	2.55%	36.05%	1.88%	3.58%	82.31%
CTTKG_2_open	59.16%	58.89%	59.02%	80.24%	79.86%	80.05%	77.41%	77.05%	77.23%	92.62%
CTTKG_2_TE	71.99%	63.33%	67.38%	71.98%	76.18%	74.02%	71.98%	72.28%	72.13%	96.90%

Figure 7 shows the tagging result of the sentence that is mentioned at the beginning of this paper. There are still some mistakes, although our model tends to find out long temporal expressions. We list some examples of annotation failures in Table 6 and classify failures roughly in three types: Wrong, Too long and Missing. The Wrong type means that some non-temporal phrases are mistaken. The most common cases are organization names, such as publishers and companies. This is mainly caused by mistakes of the knowledge graph itself. For example, the publication time of many books are incorrectly filled with publishers. This is also the reason for the Too long Tagging which mistakes words like prepositions and verbs around the temporal expression. The Missing tagging mainly occurs when the time word is short, since we only use long temporal candidates from CN-DBPedia to ensure the quality of annotation. Sometimes irregular sentences with long temporal expression which can be divided into several sub expressions also cause Missing as Fig. 7 shows.

Sentence:	1 8 9 6年农历丙申年（猴年）清光绪2 2年，孙中山在英国伦敦蒙难。
Complete tagging:	B I I I I I I I I I I I I I I I I I I I I I I O O O O O O O O O O O O
Our tagging:	B I I I I I I I I I O O O O B I I I I I O O O O O O O O O O O O

Fig. 7. Our temporal tagging result for a sentence with typical long temporal expression.

Table 6. Examples of temporal tagging failures.

Failure type	Failure examples		
	Sentence	Temporal in sentence	Mark out
Wrong	经中国证监会正式...		中国证监会
Too long	土楼始建于清代...	清代	始建于清代
Missing	后周世宗柴荣...	后周	

5   Conclusion and Future Work

In this paper, we propose a distant-supervised approach to introduce knowledge from CN-DBPedia into Chinese temporal tagging and achieve acceptable performance without any manual pattern. Especially, it performs well on open domain text and complex Chinese temporal expressions. Our two round annotation strategy combining knowledge and POS makes the model more extensible to different types of corpora. Analysis of the results shows that this method still has much to be improved. The data noise problem caused by distant supervision has yet to be better resolved. Obviously, the BiLSTM+CRF we use may not be the most suitable sequence labeling model for temporal tagging, which needs to be further explored.

References

1. Yin, B., Jin, B.: A multi-label classification method on chinese temporal expressions based on character embedding. In: International Conference on Information Science and Control Engineering, pp. 51–54. IEEE (2017)
2. Xu, B., et al.: CN-DBpedia: a never-ending Chinese knowledge extraction system. In: Benferhat, S., Tabia, K., Ali, M. (eds.) IEA/AIE 2017. LNCS (LNAI), vol. 10351, pp. 428–438. Springer, Cham (2017). https://doi.org/10.1007/978-3-319-60045-1_44
3. Walker, C., et al.: ACE 2005 multilingual training corpus. Progress Theor. Phys. Suppl. **110** (110), 261–276 (2006)
4. Verhagen, M., et al.: SemEval-2010 task 13: TempEval-2. In: Proceedings of the 5th International Workshop on Semantic Evaluation Association for Computational Linguistics, pp. 57–62 (2010)
5. UzZaman, N., et al.: SemEval-2013 task 1: TempEval-3: evaluating time expressions, events, and temporal relations. In: Second Joint Conference on Lexical and Computational Semantics (* SEM): Proceedings of the Seventh International Workshop on Semantic Evaluation (SemEval 2013), vol. 2 (2013)
6. Strötgen, J., Gertz, M.: Multilingual and cross-domain temporal tagging. Lang. Resour. Eval. **47**(2), 269–298 (2013)
7. Strötgen, J., Gertz, M.: A baseline temporal tagger for all languages. In: Conference on Empirical Methods in Natural Language Processing, pp. 541–547 (2015)
8. Li, H., et al.: Chinese temporal tagging with HeidelTime. In: Conference of the European Chapter of the Association for Computational Linguistics: Short Papers, vol. 2, pp. 133–137 (2014)
9. Angeli, G., Uszkoreit, J.: Language-independent discriminative parsing of temporal expressions. In: Meeting of the Association for Computational Linguistics, pp. 83–92 (2013)
10. Dong, C., Zhang, J., Zong, C., Hattori, M., Di, H.: Character-based LSTM-CRF with radical-level features for Chinese named entity recognition. In: Lin, C.-Y., Xue, N., Zhao, D., Huang, X., Feng, Y. (eds.) ICCPOL/NLPCC-2016. LNCS (LNAI), vol. 10102, pp. 239–250. Springer, Cham (2016). https://doi.org/10.1007/978-3-319-50496-4_20
11. Chen, T., et al.: Improving sentiment analysis via sentence type classification using BiLSTM-CRF and CNN. Expert Syst. Appl. **72**, 221–230 (2017)
12. Bollacker, K., et al.: Freebase: a collaboratively created graph database for structuring human knowledge. In: SIGMOD Conference, pp. 1247–1250 (2008)
13. Suchanek, F.M., Kasneci, G., Weikum, G.: Yago: a core of semantic knowledge. In: Proceedings of WWW, vol. 272, no. 2, pp. 697–706 ((2007))
14. Auer, S., et al.: DBPedia: a nucleus for a web of open data. In: The Semantic Web, International Semantic Web Conference, Asian Semantic Web Conference, ISWC 2007 +ASWC 2007, Busan, Korea, November, pp. 722–735. DBLP (2007)
15. Mintz, M., et al.: Distant supervision for relation extraction without labeled data. In: Joint Conference of the, Meeting of the ACL and the, International Joint Conference on Natural Language Processing of the AFNLP, pp. 1003–1011. Association for Computational Linguistics (2009)
16. Huang, B., et al.: Research on noise reduction in distant supervised personal relation extraction. Comput. Appl. Softw. **34**(7), 11–18 (2017)

17. Pustejovsky, J., et al.: The specification language TimeML. Comput. Linguist. **32**(3), 445–446 (2005)
18. Huang, Z., Xu, W., Yu, K.: Bidirectional LSTM-CRF models for sequence tagging. In: Computer Science (2015)
19. Lample, G., et al.: Neural architectures for named entity recognition, pp. 260–270 (2016)
20. Bengio, Y., Simard, P., Frasconi, P.: Learning long-term dependencies with gradient descent is difficult. IEEE Trans Neural Netw. **5**(2), 157–166 (2002)

Convolutional Neural Network-Based Question Answering Over Knowledge Base with Type Constraint

Yongrui Chen, Huiying Li[(⊠)], and Zejian Xu

School of Computer Science and Engineering, Southeast University,
Nanjing, China
18795865715@163.com, {huiyingli,220161578}@seu.edu.cn

Abstract. We propose a staged framework for question answering over a large-scale structured knowledge base. Following existing methods based on semantic parsing, our method relies on various components for solving different sub-tasks of the problem. In the first stage, we directly use the result of entity linking to obtain the topic entity in a question, and simplify the process as a semantic matching problem. We train a neural network to match questions and predicate sequences to get a rough set of candidate answer entities from the knowledge base. Unlike traditional methods, we also consider entity type as a constraint on candidate answers to remove wrong candidates from the rough set in the second stage. By applying a convolutional neural network model to match questions and predicate sequences and a type constraint to filter candidate answers, our method achieves an average F_1 measure of 74.8% on the WEBQUESTIONSSP dataset, it is competitive with state-of-the-art semantic parsing approaches.

Keywords: Question answering · Type constraint
Convolutional neural network · Knowledge base · Semantic parsing

1 Introduction

Large-scale knowledge base (KB) which is a structured database comprising vast amounts of world's facts, such as Freebase [1], DBpedia [2], have become an important resource for open-domain question answering (QA). Question answering over knowledge bases (KB-QA) is an active research area where the goal is to provide crisp answers to natural language questions [3]. One direction in KB-QA performs the answering via semantic parsing: translating natural language questions posed by humans into structured queries (e.g. SPARQL). However, the translating from question to the corresponding KB query is still a challenging task. One challenge is the *lexical gap* [4] that the relation expressed in a question can be quite different from which used in the KB. Another challenge is the *additional constraints* that need to be handled when dealing with complex questions. In this paper, we handle these problems with a staged framework.

Our approach has two stages. In the first stage, after obtaining the *topic entity* of the question, we detect the relation asked by the question for all the relations (relation or

© Springer Nature Singapore Pte Ltd. 2019
J. Zhao et al. (Eds.): CCKS 2018, CCIS 957, pp. 28–39, 2019.
https://doi.org/10.1007/978-981-13-3146-6_3

relation chain with limited length) that starts from the topic entity. The candidate answer entities can be retrieved with the topic entity and the detected relation. In the second stage, we add a *type constraint* to the candidate answer entities. We predict the answer entity types to remove wrong answers and obtain more accurate final answers. Inspired by [5], we employ the neural network models to handle the relation detection and type prediction in our approach.

It is a significant difference from previous methods that we consider the constraint of the entity type. We notice that a wrong answer usually has an incorrect type. For example, given question "What highschool did Harper Lee go to?", five candidate answer entities can be retrieved along the relation chain "education-institution" in the first stage of our approach. However, only the entity "Monroe County High School" with type "School" is the correct answer, other candidates' type "College/University" are wrong. For improving the accuracy of results, we add type constraint on the candidate answer entities. Our question answering system improves the state-of-the-art result of on the WEBQUESTIONSSP dataset [6] to 74.8% in F1, a 3.1% absolute gain compared to the best existing method.

The rest of this paper is structured as follows. Section 2 introduces the overview of related work for KB-QA. Section 3 presents our staged approach based on CNN. The experimental results are shown in Sect. 4. Finally, Sect. 5 concludes the paper.

2 Related Work

An important line of research for KB-QA constructs an end-to-end system by deep learning powered similarity matching. The aim of these approaches is to learn semantic representations of both the question and the KB elements with an encoder-compare framework, so that the correct KB element is the nearest neighbor of the question in the learned vector space [7]. These approaches are different on the used neural network models, such as CNN or RNN, and the input granularity, such as word-level or character-level. Denis et al. [4] concentrate two granularities of relation. They propose a GRU-based model on both word and character level. Yu et al. [8] propose a hierarchical RNN enhanced by residual learning, which detects different levels of abstraction. Qu et al. [9] propose an attentive RNNs with similarity matrix based CNNs to preserve more original words interaction information. These end-to-end methods usually have only a single process in order to avoid complex NLP pipeline constructions and error propagation. Another advantage is they can be retrained or reused for a different domain [4]. However, the obvious drawback of these methods is the results are often unexplainable.

Another important line of research in this domain performs answering based on semantic parsing. The core idea of semantic parsing is to translate a question into a structured KB query. Then, the answer to the question can be retrieved simply by executing the query. Early semantic parsing approaches require complex NLP pipeline components, such as POS tagger, template-fitting, relation-extraction. Berant et al. [10] train a semantic parser that scales up to Freebase with question-answer pairs instead of annotated logical forms. Reddy et al. [11] represent natural language via semantic graphs. They conceptualize semantic parsing as a graph matching problem. Fader et al.

[12] propose an open question answering over curated and extracted knowledge bases. Berant et al. [13] also propose an approach that turns semantic parsing on its head via paraphrasing. With the maturity of deep learning, recent work has focused on building semantic parsing frameworks with neural networks to improve the effect. Xu et al. [14] propose SDP-LSTM to classify the relation of two entities in a sentence. Yih et al. [5] simplify the process of sematic parsing as the generation of the query graph. They train a CNN for the representation of the question and the relation. Dong and Lapata [15] propose an attention-enhanced encoder-decoder model to generate the logic form for the input utterance based on LSTM. Comparing with the end-to-end methods, semantic parsing can provide a deeper understanding of the question [4]. Moreover, process of answering which is staged can be easily for error analysis. In contrast, semantic parsing approaches usually have the disadvantage of error propagation and complicated steps.

Our approach is closely related to the second line of research with semantic parsing but simplifies the complicate steps as a semantic matching problem. Different from previous methods, we apply a type constraint to filter candidate answer entities.

3   Approach

We focus on the simple question that contains only one topic entity in this work. The problem is solved by the following two stages:

1. Mapping the words sequence in a question to a relation (or a relation chain with limited length, we use relation to refer both relation and relation chain for brevity) in the KB, it is also referred to *relation detection*. We reduce this problem to measuring semantic similarity of the words sequence and the relation.
2. Adding constraint. Inspired by [18], we find that answer type is also an important aspect for understanding question. Therefore, we take the entity type as a constraint to filter the correct answer entities from all candidates found by the first stage. We consider the process of *answer entity type prediction* as a binary classification problem.

For handling the multiple equivalent semantic statements of the same question, as well as the lexical gap between the natural language utterance and relation in the KB, we propose using Convolutional Neural Networks (CNN) [16, 17, 19] for relation detection and answer entity type prediction. The model is further elaborated in the next section (Sect. 3.1), followed by a description of the training process (Sect. 3.2).

3.1   Model Description

As illustrated in Fig. 1, under the precondition that the topic entity is identified by entity linking, our model consists of three components:

1. Two CNNs to produce low dimensional vector representations for the question (CNN_q) and the relation (CNN_r). For all relations starting from the topic entity in the KB, we compare its vector representation with the representation of the question to detect the relation asked by the question (Fig. 1a).

2. A CNN to predict answer entity types. Based on the topic entity and the detected relation, we form a query to search the KB for candidate answer entities. For all candidate types, we use CNN_t to produce 2-dimensional vectors of question-type pairs for answer entity type prediction (Fig. 1c).
3. The process of retrieving answers. We apply the predicted types as a constraint to filter correct answer entities from all candidates (Fig. 1b).

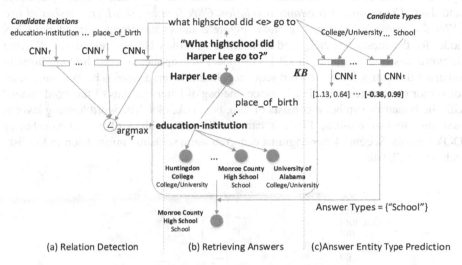

(a) Relation Detection   (b) Retrieving Answers   (c)Answer Entity Type Prediction

Fig. 1. The overview of the proposed approach.

The details of each component are described in the following paragraphs.

Retrieving Answers. The process of retrieving answers (Fig. 1b) relies on relation detection and answer entity type prediction. Given a question q and the corresponding topic entity e_t, we present the question by replacing mention of e_t with symbol "<e>" for noise reduction. Then, our question answering approach proceeds in the following two stages:

1. Employ a relation CNN framework to detect the relation r for question q, and retrieve the candidate answer entities that has relation r connecting e_t in the KB. The candidate answer entities set is denoted as C_a^q.
2. Generate an answer entity type set T_a of q with the type prediction of another CNN framework, and remove the entities from C_a^q whose type is not in T_a to obtain the final answer entity set A_f.

Relation Detection. We construct an encoder-compare framework for relation detection. The process can be described as three steps as followed:

1. Generate the relation candidates set C_r^q for question q. We select all relations and chains of relations starting from the topic entity e_t. For a chain of relations, we limit its length to 2.

2. Extract k-dimentional semantic vectors for both q and each candidate relation in C_r^q by the relation CNN.
3. Calculate and then compare the similarity score between each candidate relation in C_r^q and q and take the top-scored one as the detected relation r.

The first step generates C_r^q including the relation chains which start from the topic entity. In Freebase's design, there is a special entity category called compound value type (CVT). Therefore, we consider the relation and relation chains (length ≤ 2) as the candidate. *The second step adapts a relation CNN framework which consists of two CNNs for extracting semantic vectors.* Figure 2 shows the architecture of the CNN model for the question (CNN$_q$) and it is similar with the CNN$_r$ for relations. Following the word hashing technique in [20], the model first applies a word hashing layer to construct the matrix for input word sequence. In this layer, a word is broken into a one-hot vector of letter-trigrams. For instance, the bag of letter-trigrams of the word "book" with the boundary symbol # consists of #bo, boo, ook, ok#. The word hashing layer is used due to two reasons. First, it can handle the problem of out of vocabulary (OOV) words. Second, letter-trigrams can grab some semantic information in English, such as "pre", "dis".

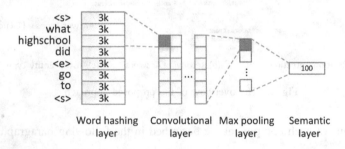

Fig. 2. The architecture of the CNN used for question to produce the vector.

After word hashing, the model use k kernels with a window size of three words to project the word sequence matrix to k vectors which hold the local features. Then, the max pooling layer extracts a global feature vector of length k by performing the maximum operation on each local feature vector. The following semantic (fully-connected) layer projects the global feature vector to a l-dimensional semantic vector of the question. We employ the CNN because its convolutional and max pooling layer can extract both local and global semantic features of word sequence, and make our model has more representation power.

The third step detects the relation asked by the question. With the semantic vectors of both question q and candidate relation c_r, we compute the similarity score between p and c_r as follows:

$$S(q, c_r) = \cos\left(v_q, v_{c_r}\right) \tag{1}$$

where v_q and v_{c_r} are the semantic vectors of the q and c_r respectively, cos is the cosine similarity.

Answer Entity Type Prediction. Although predicting answer entity types is still a matching problem of question and answer type, it different from relation detection. Because one question only corresponds to one relation but can correspond to more than one answer type. For instance, for question "where is the fukushima daiichi nuclear plant located?" the two answer entities Japan and Okuma have different types "Country" and "City/Town/Village". Therefore, we consider the matching problem as a task of binary classification for each question-type pair, which is proceeded as the following three steps.

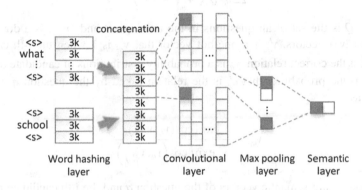

Fig. 3. The architecture of the type CNN.

The first step generates a type candidates C_t^q consists of type labels for the question q. We collect the lowercased English type labels (for Freebase, we use the **common.topic.notable_types** property) for each candidate answer entity generated by the topic entity and the detected relation from the KB, and take them as candidate types.

The second step concatenates q and each type in C_t^q to a new word sequence and feed it into the type CNN to output a prediction vector indicates whether the candidate type is positive. The type CNN framework is illustrated in Fig. 3. Different from the CNN for relation detection, we concatenate the sequence matrix of question and type and feed it to the type CNN. Moreover, the type CNN uses the kernels with two kinds of window sizes, three words and four words respectively to extract local features. The output of CNN is a 2-dimensional vector $v = [v_0, v_1]$ which is the probability distributions of the matching. The candidate type of the question is judged as positive when v_1 is greater than v_0.

The third step generates an answer entity type set T_a which consists of all positive types in C_t^q.

3.2 Training

We treat relation detection as a multi-class classification problem while answer entity types prediction as a binary classification problem. So we train the relation CNN and the type CNN in different ways. We detail the loss functions and negative sampling strategies in the next few paragraphs.

Loss Function. For the matching of question and relation, we drive the model to output a higher similarity score for the correct pair than wrong pairs. We consider each candidate relation as a class and use the cross entropy loss function minimized during training is given by

$$\sum_{q \in Q} \left\langle y'_q, \ \log(y_q) \right\rangle \tag{2}$$

That is, Q is the set of all questions in the training set and $\langle x, y \rangle$ is a dot product operation of two vectors. y'_q is a one-hot vector that y'_{qi} is 1 when the i-th candidate relations c^i_r is the correct relation. y_q is a probability distributions of candidate relations, that is, y_{q_i} is the probability that c^i_r is the relation asked by the question q, which is calculated as

$$y_{q_i} = \frac{\exp\left(\cos\left(v_q, v_{c^i_r}\right)\right)}{\sum \exp\left(\cos\left(v_q, v_{c^j_r}\right)\right)} \tag{3}$$

where v_q and $v_{c^i_r}$ are semantic vectors of the question q and the i-th candidate relations c^i_r respectively.

For predicting the answer types, we project the concatenated word sequence for each question-type pair to a 2-dimensional vector v by the type CNN. Then, we normalize v to get a probability distributions y_{pt} by the softmax function. The loss function for training pattern-type pairs is given by

$$\sum_{q \in Q, t \in C^q_t} \left\langle y'_{q,t}, \ \log(y_{q,t}) \right\rangle \tag{4}$$

where $y'_{q,t}$ is a 2-dimensional vector that indicates whether t is a positive type or not.

Negative Sampling. For training CNN, the positive relation sample s^+_r, which is the correct relation r of a question q, can be obtained from the sematic parses provided in the training data. We can also collect the correct answer entity types t_a as positive type samples set T^+_a of a question q from the training set. However, the dataset does not provide the negative samples for training. Therefore, we need perform negative sampling from the KB. The generation of corrupted samples are based on s^+_r and T^+_a. We observe that the way of negative sampling has a significant influence on results. In principle, we try to simulate a condition of training which is close to the condition during the test.

Relation Corruption. To corrupt s_r^+, a false relation can be sampled from the space of all relations in the KB. However, we tend to make the model more distinguishable from more plausible wrong relations. We use the same candidate relation generation procedure as the one used for test. The negative relation sample set N_s is generated as

$$N_s = C_r^q - \{s_r^+\} \tag{5}$$

where C_r^q is the candidate relation set for a question q in the training set.

Type Corruption. We collect the type label of Freebase entity by its **common.topic. notable_types** relation. To make the model more distinguishable, we train the type CNN to predict correct answer entity types from the candidate types C_t^q. The negative type sample set N_t is generated as

$$N_t = C_t^q - T_a^+ \tag{6}$$

where C_t^q is a type set of all the entities in C_a^q. These entities are retrieved along the relation s_p^+ from the topic entity in the KB.

In our experiments, we observed that the number of negative type samples is about one third of the number of the positive type samples. The unbalance in quantity between positive and negative samples leads to the result that the model tends to judge negative sample as a positive one. So, we adapt a simple over-sampling strategy that training three times on N_t while one time on T_a^+ at each iteration. Then, we achieve a significant better result on answer entity type prediction than before.

4 Experiments

In this section, we detail the experiments we made to evaluate our model. First we introduce the dataset and evaluation metric, followed by the main experimental results and error analysis.

4.1 Data and Evaluation Metric

We evaluate the proposed approach on the WEBQUESTIONSSP dataset [6], which contains full semantic parses for a subset of the questions from WEBQUESTIONS dataset [10]. It removes 18.5% of the origin dataset because these questions are considered as "not answerable". The dataset consists of 4737 question-answer pairs which are split into training set of 3098 and test set of 1639. The questions of dataset were collected using Google Suggest API and the answers were obtained using Amazon Mechanical Turk workers. They were updated in [6] using Freebase and added full semantic parses.

In our experiments, we directly obtain the topic entity of each question by the entity linking results from the semantic parses provided in the dataset for both training and testing. In addition, we drop some questions which have no answers or no relation during training, and out model outputs empty answer entity sets for this kind of questions on the evaluation stage. The system performance is basically measured by the ratio of questions that are answered correctly. Because there can be more than one answer to a question, *precision*, *recall*, and F_1 metrics are computed based on the output of each individual question, and average F_1 is reported as the main evaluation metric. We compute experiment results with the official evaluation script[1].

4.2 Experimental Settings

We add boundary symbols "<s>" at the beginning and end of all word sequences. In order to solve the difference in word sequence length, we specify that each question has a length of 20 words, each candidate relation has a length of 4 words, each question-type pair has a length of 25 words. For word sequences of insufficient length, we add all-zero letter-trigrams vectors at the end of the word hashing matrix to reach the specified length.

Because the CNNs of our model are used for different tasks, we do not set the hyper-parameters for training them in the same way. For the relation CNN, the convolutional layer size is set to 10000, the max pooling layer size is set to 500, the semantic (output) layer size is set to 100, the learning rate is 0.01 and the training process lasts 200 epochs. For the type CNN, the convolutional layer size is set to 20000, the max pooling layer size is set to 2000, the semantic (output) layer size is set to 2, the learning rate is 0.001 and the training process lasts 200 epochs.

Table 1. Comparison of our approach with existing work.

Method	Prec.%	Rec.%	Avg. F_1%
Yih et al. 2016 (Answers) [6]	67.3	73.1	66.8
Liang et al. 2017 [21]	70.8	76.0	69.0
Yih et al. 2016 (Sem. Parses) [6]	70.9	80.3	71.7
Our approach (Rel. Detection)	68.9	82.4	70.8
Our approach (Rel. Detection + Type Prediction)	79.9	82.3	**74.8**

4.3 Results

Table 1 shows the overall results of our approach compared to existing work [6, 21] on the WEBQUESTIONSSP dataset. The results show that our approach outperforms the previous state-of-the-art method by 3.1% in average of F_1 measure.

The main difference of our approach comparing with previous methods is that we use the type constraint on candidate answer entities to improve the precision. Therefore, for examining the contribution of the type constraint, we test the performances with and without the type constraint in Table 1. When the retrieval of answers relies only on the relation detection, the performance of our system is close but does not exceed previous work. After adding the predicted entity type to constrain candidate

---

[1] Available at http://aka.ms/WebQSP.

answers, the performance of the whole approach increases 4% in F_1. The 10% increase in precision indicates that our type constraint is effective on removing wrong answers.

For the perspective of the neural networks, the relation CNN achieves 79.7% accuracy on relation detection, and the type CNN achieves 91.1% in $F - 1$ (Prec. = 90.5%, Rec. = 94.1%) on predicting answer entity types for all questions in the test data. For examining the contribution of the word hashing layer in CNN, we test the performances of the relation CNN on word-level and character-level. On word-level, we use *GloVe* [22] vectors provided on the Glove website[2] as word embeddings. On character-level, we first consider both question and relation as character sequences. Then, we use the character vectors in Glove to construct the matrix for input character sequence. The results are shown in Table 2. Both word and character-level are far less effective than word hashing technique on the proposed relation CNN.

Table 2. The performances of the relation CNN on word-level and character-level.

Granularity	Acc. %
Word-level	52.3
Character-level	55.8
Word-hashing	79.7

4.4 Error Analysis

Although our approach substantially outperforms existing methods, it is partly because we directly use the results of entity linking provided in the dataset and there is no error on obtaining topic entity.

For error analysis, we randomly sampled 100 questions from 500 questions which our system did not output the completely correct answer set. We find that 57% of the errors are due to the incorrect relation detection, 23% are due to ignoring the time constraints of some questions. For example, for question "What team does Jeremy Lin play for *2013*?" we output all teams that Jeremy Lin played for without considering the time constraint. We don't consider other similar time constraints either, such as "first", "now". Existing more than one entity in a question like "who plays *Lex Luthor* on *Smallville*?" also causes 8% of the errors. 3% of the errors are due to the types used for constraint are too wide to remove wrong answers. For instance, the answer entity type we predicted for question "What is the state flower of New Mexico?" is "Official Symbol". The rest 9% errors are caused that the answers retrieved from the KB are not completely same with the gold answers when all stages are correct.

---

[2] https://nlp.stanford.edu/projects/glove/.

5 Conclusion

In this paper, we present a staged, convolutional neural network-based approach with type constraint for question answering over knowledge base. In the first stage, we detect the relation that starts from the topic entity to obtain candidate answer entities. We also add the type constraint on candidate answer entities to remove wrong answer entities in the second stage. With the help of the convolutional neural network model and topic entity linking results provided in the dataset, our approach outperforms the previous method on the WEBQUESTIONSSP dataset.

Acknowledgements. The work is supported by the Natural Science Foundation of China under grant No. 61502095, and the Natural Science Foundation of Jiangsu Province under Grant BK20140643.

References

1. Bollacker, K., Evans, C., Paritosh, P., et al.: Freebase: a collaboratively created graph database for structuring human knowledge. In: Proceedings of the 2008 ACM SIGMOD International Conference on Management of Data, pp 1247–1250. ACM, New York (2008)
2. Auer, S., Bizer, C., Kobilarov, G., Lehmann, J., Cyganiak, R., Ives, Z.: DBpedia: a nucleus for a web of open data. In: Aberer, K., et al. (eds.) ASWC/ISWC -2007. LNCS, vol. 4825, pp. 722–735. Springer, Heidelberg (2007). https://doi.org/10.1007/978-3-540-76298-0_52
3. Xiao, Y., Wang, H., Song, Y., Hwang, S., Wang, W.: KBQA: learning question answering over QA corpora and knowledge bases. In: Proceedings of the 42nd International Conference on Very Large Data Bases, New Delhi, India, pp. 565–575. ACM (2016)
4. Denis, L., Asja, F., et al.: Neural network-based question answering over knowledge graphs on word and character level. In: 26th International World Wide Web Conference, Perth, Australia, pp. 1211–1220 (2017)
5. Yih, S.W.T., Chang, M.W., et al.: Semantic parsing via staged query graph generation: question answering with knowledge base. In: Proceedings of the 53rd Annual Meeting of the Association for Computational Linguistics, Beijing, China, pp. 1321–1331. Association for Computational Linguistics (2015)
6. Yih, W., Richardson, M., Meek, C., et al.: The value of semantic parse labeling for knowledge base question answering. In: Proceedings of the 54th Annual Meeting of the Association for Computational Linguistics, Berlin, Germany, pp. 201–206. Association for Computational Linguistics (2016)
7. Dai, Z., Li, L., Xu, W.: CFO: conditional focused neural question answering with large scale knowledge bases. In: Proceedings of the 54th Annual Meeting of the Association for Computational Linguistics, Berlin, Germany, pp. 800–810. Association for Computational Linguistics (2016)
8. Yu, M.: Improved neural relation detection for knowledge base question answering. In: Proceedings of the 55th Annual Meeting of the Association for Computational Linguistics, Vancouver, BC, Canada, pp. 571–581. Association for Computational Linguistics (2017)
9. Qu, Y., Liu, J., Kang, L., et al.: Question answering over freebase via attentive RNN with similarity matrix based CNN. arXiv preprint arXiv:1804.03317 (2018)

10. Berant, J., Chou, A., Frostig, R., Liang, P.: Semantic parsing on freebase from question-answer pairs. In: Proceedings of the 2013 Conference on Empirical Methods in Natural Language Processing, Seattle, Washington, USA, pp. 1533–1544. Association for Computational Linguistics (2013)

11. Reddy, S., Lapata, M., Steedman, M.: Large-scale semantic parsing without question-answer pairs. Trans. Assoc. Comput. Linguist. 2(2014), 377–394 (2014)

12. Fader, A., Zettlemoyer, L., Etzioni, O.: Open question answering over curated and extracted knowledge bases. In: Proceedings of the 20th ACM SIGKDD International Conference on Knowledge Discovery and Data Mining. pp. 1156–1165. ACM, New York (2014)

13. Berant, J., Liang, P.: Semantic parsing via paraphrasing. In: Proceedings of the 52nd Annual Meeting of the Association for Computational Linguistics, Baltimore, Maryland, pp. 1415–1425. Association for Computational Linguistics (2014)

14. Yan, X., Mou, L., Li, G., et al.: Classifying relations via long short term memory networks along shortest dependency path. In: The 2015 Conference on Empirical Methods in Natural Language Processing, Lisbon, Portugal, pp. 1785–1794. Association for Computational Linguistics (2015)

15. Dong, L., Lapata, M.: Language to logical form with neural attention. In: Proceedings of the 54th Annual Meeting of the Association for Computational Linguistics, Berlin, Germany, pp. 33–43. Association for Computational Linguistics (2016)

16. Zeng, D., Liu, K., Lai, S., et al.: Relation classification via convolutional deep neural network. In: The 25th International Conference on Computational Linguistics: Technical Papers, Dublin, Ireland, pp. 2335–2344. ACM (2014)

17. Shen, Y., He, X., Gao, J., Deng, L., Mesnil, G.: A latent semantic model with convolutional-pooling structure for information retrieval. In: Proceedings of the 23rd ACM International Conference on Conference on Information and Knowledge Management, Shanghai, China, pp. 101–110. ACM (2014)

18. Dong, L., Wei, F., Zhou, M., Xu, K.: Question answering over freebase with multi-column convolutional neural networks. In: Proceedings of the 53rd Annual Meeting of the Association for Computational Linguistics and the 7th International Joint Conference on Natural Language Processing, Beijing, China, pp 260–269 (2015)

19. Gao, J., Pantel, P., Gamon, M., He, X., Deng, L., Shen, Y.: Modeling interestingness with deep neural networks. In: Proceedings of the 2013 Conference on Empirical Methods in Natural Language Processing (2014)

20. Huang, P., He, X., Gao, J., Deng, L., Acero, A., Heck, L.: Learning deep structured semantic models for web search using clickthrough data. In Proceedings of the 22nd ACM International Conference on Conference on information and knowledge management, San Francisco, CA, USA, pp. 2333–2338. ACM (2013)

21. Liang, C., Beranty, J., Le, Q., Forbus, K.D., Lao, N.: Neural symbolic machines: learning semantic parsers on freebase with weak supervision. In: Proceedings of the 55th Annual Meeting of the Association for Computational Linguistics, Vancouver, BC, Canada, pp. 23–33. Association for Computational Linguistics (2017)

22. Pennington, J., Socher, R., Manning, C.D.: Glove: global vectors for word representation. In: Proceedings of the 2014 Conference on Empirical Methods in Natural Language Processing, Doha, Qatar, pp 1532–1543. Association for Computational Linguistics (2014)

MMCRD: An Effective Algorithm for Deploying Monitoring Point on Social Network

Zehao Guo, Zhenyu Wang[✉], and Rui Zhang

South China University of Technology, Guangdong 510006, China
marcozhguo@gmail.com, wangzy@scut.edu.cn, z.rui16@mail.scut.edu.cn

Abstract. Complex relationships and restrictions on social networking sites are severe issues in social network data acquisition. Covering information of all users in social network and ensuring timeliness of data acquisition is of great significance. Therefore, it is critical to develop an efficient data acquisition strategy. In particular, smart deployment of monitoring points on social networks has a great impact on data acquisition efficiency. In this paper, we formulate the monitoring point deployment issue as a capacitated set cover problem (CSCP) and present a maximum monitoring contribution rate deployment algorithm (MMCRD). We further compare the proposed algorithm with random approximation deployment algorithm (RD) and maximum out-degree approximation deployment algorithm (MOD), using synthetic BA scale-free networks and real-world social network datasets derived from Facebook, Twitter and Weibo. The results show that our MMCRD algorithm is superior to the other two deployment algorithms, since our approach can monitor the entire social network users by monitoring at most 12% of users, and meanwhile, guarantee timeliness.

Keywords: Data acquisition · Timeliness
Monitoring point deployment · Capacitated set cover problem
MMCRD

1 Introduction

Social networking sites (e.g., Facebook, Twitter and Weibo) have emerged as main channel for people to share and exchange information on a wide range of real-world events. These events range from popular and well-known ones (e.g., a concert of a famous music band) to a small-scale local event (e.g., a local gathering and protest). Short messages posted on social networking sites reflect events that are taking place and spread in the form of "nuclear fission" through relationship network. Monitoring real-time messages in social network has an important significance in burst event detection [1], rumor detection [2], natural disaster detection [3], event topic identification [4], stock markets forecasting [5] and other fields.

© Springer Nature Singapore Pte Ltd. 2019
J. Zhao et al. (Eds.): CCKS 2018, CCIS 957, pp. 40–51, 2019.
https://doi.org/10.1007/978-981-13-3146-6_4

Three technologies are commonly adopted to acquire social networking data, namely application programming interface (API), asynchronous loading based on communication simulation and reverse crack engineering. API is restricted by resources and has strict limits on the frequency, authority and content of call interfaces. Asynchronous loading need to overcome a variety of measures against crawler and analyze the perplexing dependencies among requests, which is difficult to acquire complete data. Reverse crack engineering requires a huge and unmanageable reverse engineering, the escalating security measures make the cost of crack engineering increase and the risk is difficult to control.

This paper use API to acquire social networking data. Two key requirements of acquiring social networking data are a full coverage of users in network and timely data access as data are generated on the networks continuously.

To meet the first requirement, in traditional ways, we need to invoke API periodically for each user on social network, which will certainly require massive API calls due to the sheer number of users. To meet the second requirement, we require to monitor users as less as possible. Obviously, with given amount of resources, we have to develop an effective strategy to deploy monitoring points on social network.

Based on the analysis of user relationships and information dissemination in social network, we formulate monitoring point deployment issue as a CSCP and propose MMCRD. We further compare the proposed algorithm with RD and MOD in terms of the relationship between monitor capacity and required number of monitors by using real-world datasets derived from Facebook, Twitter and Weibo and synthetic BA scale-free network. The results show MMCRD is superior to the other two deployment algorithms, we only need to monitor at most 12% users in order to achieve full coverage and guarantee timeliness by using MMCRD.

2   Related Work

Online social network is a social structure which is formed by the individuals and relationships between the individuals. The relationships between individuals can be friends, relatives, behavior interactions and other relationships. Active behaviors of individuals in social network promote the constant changes of network structure, which makes social network hold the characteristics of community interaction [6], information dissemination [7] and community evolution [8]. Social networks have a large volume of valuable information. Compared to crawl news sites, blogs, forums and peer-to-peer systems, acquiring social networking data is really a difficult job. Social networks consist of three types of data, namely user profiles, user relationships and user-generated contents (tweets). User profiles and user relationships are relatively stable while user-generated contents change constantly.

In recent years, a great deal of research work acquire social networking data by means of traversing user graph which is feasible in most social networking sites, including Facebook [9,10], Ebay [11,12], Youtube [13,14], Flickr [15,16],

Twitter [17,18] and Weibo [19]. User graph crawling is an iteration process that starts from one or more seed users and depends on user relationships to discover new user in each iteration. It can exploit the skeleton of social network and easy to acquire user profiles and relationships. In addition, several strategies are commonly adopted to traverse user graph, including Bread-First-Search, Depth-First-Search, Forest Fire, Snowball Sampling and Respondent-Driven Sampling, they are differ in order through which they visit new users in each iteration.

Crawling is the most frequently used solution for data acquisition and is effective indeed to obtain user profiles and user relationships whereas is not suitable for acquiring user-generated contents. In most cases, acquiring user-generated contents is based on acquired users that means we can not satisfy the requirement of full coverage as long as there is no acquisition for some users. Behind the difficulty mentioned above is mainly because of (1) huge size of user relationship graph. (2) restrictions on social networking sites, e.g., IP banning.

Due to the above-mentioned problems, it is real a challenge to acquire the constantly update of user-generated contents timely. The purpose of this paper is to propose an effective monitoring point deployment algorithm to achieve full coverage of users and timely acquisition of user-generated contents.

3   Problem Formulation

3.1   Deploy Monitoring Points on Social Network

Four main behaviors are included in social network, namely *follow*, *post*, *retweet* and *mention*. The *follow* behavior builds a following relationship which require no mutual approval and is often regarded as a form of subscription. Other behaviors (*post*, *retweet* and *mention*) are relevant to user-generated contents. The relationships among users are built based on *follow* behaviors. If user A follows user B, user A is called a follower of user B, conversely, user B is called a friend of user A. In such a situation, if user B posts a new tweet (or retweet), this new tweet (or retweet) will appear in the main page of user A. In other words, if user A follow user B, information generated by user B will diffuse to user A. The following relationships and information diffusion can be shown as Fig. 1. The solid lines denote following relationships and the dash lines denote information diffusion directions. For example, user A follow user B and user C, the information generated from user B and user C will disseminate to user A. Therefore, as long as we deploy a monitor on user A, we can obtain all the information generated from user A, B and C. Other users are the same situation. Therefore, if we put monitors on user A, D, E, F, we can acquire the full information generated from all users in Fig. 1.

Based on user following relationships and information diffusion process, we can deploy monitoring points on a set of users: $U = \{u_1, u_2, ..., u_m\}$, and let each user u_i to cover a group of users, for example, the group of users covered by u_i can be represented as $V_i = \{v_{i1}, v_{i2}, ..., v_{ij}\}$. If

$$\bigcup_{i=1}^{m} V_i = \Omega \tag{1}$$

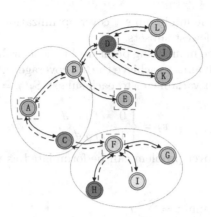

Fig. 1. User following relationships and information diffusion (The solid lines denote user following relationships; The dash lines denote information diffusion directions).

where Ω represents the whole users on a social network. Then we can monitor the set of users U to achieve full coverage of the users in social network.

Definition 1. We use a directed Graph $G = (V, E)$ to represent a social network compose of users and user relationships. Let $V = \{v_1, v_2, ..., v_n\}$ represent users in social network and $E = \{e_{1,1}, e_{1,2}, ..., e_{n,n-1}\}$ represent user relationships. For example, if exists a relationship between user v_i and user v_j, it represents that user v_i follow user v_j and also represents that $e_{i,j}$ contains in set E. Based on following relationships, if exists a subset of vertices $V' \subseteq V$, for each user $v_i \in V'$, we can obtain its following set which can be represented as F_i, Let S_i represent v_i combining with F_i. If any user belong to V, it can be covered by at least one subset S_i, then we call V' is the monitor set of graph G. If the number of elements of V' is minimum, V' is called a minimum set cover of graph G.

In this way, we formulate monitoring point deployment problem as a capacitated set cover problem. Next we will discuss how to determine appropriate monitoring points to achieve full coverage of the users in network and guarantee timeliness.

3.2   Set Cover Problem Formulation

Assume we have a set $U = \{v_1, v_2, ..., v_n\}$ and m subsets $S = \{S_1, S_2, ..., S_m\}$, where $S_j \subseteq U$, $j \in J$ and $J \subseteq \{1, 2, ..., m\}$. If the following equation can be satisfied,

$$\bigcup_{j \in J}^{m} S_j = U \tag{2}$$

where we call $\bigcup_{j \in J}^{m} S_j$ the set cover of U. A set cover the problem of selecting a collection of subsets satisfied that every element in U belongs to at least one of the subset $S_j(j \in J)$. The minimum set cover optimization problem is to find a set cover that uses the fewest subsets.

We can transform the monitoring point deployment problem into a minimum set cover problem, which aims to achieve the full coverage with minimum subsets. We introduce a decision variable x_j to each subset $S_j(j = 1, 2, ..., m)$, which is described as below.

$$x_j = \begin{cases} 0 & if\ j \in J \\ 1 & if\ j \notin J \end{cases} \tag{3}$$

The minimum set cover problem can be formulated as the following integer program.

$$\min\ z = \sum_{i=1}^{m} x_i$$
$$s.t.\quad \sum_{j:e_i \in S_j} x_j \geq 1 \quad i = 1, 2, ..., n \tag{4}$$
$$x_j = 0, 1 \quad i = 1, 2, ..., n$$

The constrain $\sum_{j:e_i \in S_j} x_j \geq 1$ ensures each element e_i of S can be covered by a subset S_j.

In addition to the full coverage, monitoring point deployment also require timeliness. The delay of data acquisition is determined by the number of elements covered by each vertex. Due to information dissemination among users in the form of tweet, we need to consider tweet frequency in data acquisition. Assume the maximum tweet frequency is f tweets per second, data acquisition rate is p tweets per second and tolerable delay is t seconds, the number of elements in each subset $S_j(j \in J) : K(S_j)$ should satisfy the inequality as below.

$$k_v \leq \frac{pt}{f} \tag{5}$$

Because we need to deploy monitoring points to achieve full coverage and guarantee timeliness, in fact, is a capacitated set cover problem which is described as follow.

$$\min\ z = \sum_{i=1}^{m} x_i$$
$$\sum_{j:e_i \in S_j} x_j \geq 1 \quad i = 1, 2, ..., n$$
$$s.t.\quad x_j = 0, 1 \quad i = 1, 2, ..., n \tag{6}$$
$$|K(S_j)| \leq \frac{pt}{f}$$

where $K(S_j)$ represents the number of elements of subset $S_j(j \in J)$.

4   MMCRD for Monitoring Point Deployment

Based on the above discussion, the MMCRD algorithm is summarized as Algorithm 1.

Algorithm 1. Maximum Monitoring Contribution Rate Deployment Algorithm (MMCRD)

Input:

 vertex set: V=$v_1, v_2, ..., v_n$ adjacent matrix A; maximum tweet frequency: f tweets per second; data acquisition rate: p tweets per second; tolerable delay: t seconds;

Output: Monitoring Points Set MPS;

Steps:

 Step 1. Initialization

 Monitoring set M=Φ; Following set F=Φ;Monitored set S=Φ; Isolated set I=Φ; New Monitoring set NW=Φ;

 Step 2. Determine out-degree k_i^{out} and excess average in-degree ek_i^{in} of v_i(i = 1, 2.., n) and then calculate monitoring contribution rate MCR_i(MCR_i = $k_i^{out}/(ek_i^{in} + \varepsilon)$), determine the vertex v_j with maximum MCR;

 Step 3. M=M+$\{v_j\}$;

 Step 4. F=Φ; Find the following set of v_j and refresh F;

 Step 5. Determine the size of F represented by p;

 Step 6. $if p <= r = \frac{pt}{f}$,S=S+$\{v_j\}$+F;

   $else$ F=choose r vertices with lowest in-degree;

    S=S+v_j+F;

   end;

 Step 7. Delete users from M, S and all the related edges and refresh A;

 Step 8. if the element of A are not equal to 0,

    Then jump to step 2;

   $else$ I = V-S;

   end;

 Step 9. While $\|I\| > r$

   Choose r users from I and form a set $N=r$ users;

   Create a new user v_k to follow users in set N;

   NM = NM+$\{v_k\}$;

   I = I-N;

   end;

 Step 10. Create a new user v_k to follow $\|I\|$ users;

   NM = NM+$\{v_k\}$;

 return MPS = M+NM;

  The algorithm takes a vertex set V and an adjacent matrix A of a directed graph G as inputs and return a monitoring point set MPS. The algorithm works as follows: At the beginning, we initial the monitoring set M equal to empty set, the elements of M represent the monitoring points in graph G. The elements of the monitored set S indicate the vertices have been monitored. The role of the following set F is to temporarily store the friends of a selected monitoring point. The isolated set I contains isolated nodes which are not yet monitored by monitoring points. Referred to isolated nodes, we create new vertices to cover them, these new vertices form the new monitoring set NW. In each loop, we determine the vertex v_j with maximum monitoring contribution rate (MCR) in graph G and adds it to M. Subsequently, we find the friends of v_j and temporarily store in F. If the size of F is large than timeliness constrain ($r = pt/f$), replace F with r vertices with lowest in-degree in F, then add v_j and F to S. Loop continues until the adjacent matrix A is all 0. When the loop ends, we obtain the isolated set I by calculating the difference set of V and S. Different from step 6, for each new created vertex, we randomly choose r vertices in I to be monitored rather than choose r lowest in-degree ones for the reason that the connectivity of the remaining isolated vertices is relatively sparse and the in-degree of each vertex is very low and nearly the same. Finally, we combine the M and NW and return the monitoring point set MPS.

Figure 2 show the steps of the MMCRD algorithm using a sample graph. Figure 2(a) shows an input directed graph containing 7 vertices and 9 edges, the MCR of each vertex is initialized according to its out-degree and average in-degree of its friend(s), for example, the out-degree of C is 2, the in-degrees of its friends B and D are 2 and 4 respectively, then we can calculate the average in-degree of B and D, is equal to 3, therefore the MCR of C is 2/3. Other vertices are the same situation. The vertex with highest MCR is B, so it is added to M, both B and its friend A are added to S as shown in Fig. 2(b). B and its friend are discolored and all edges incident on them are dashed. Figure 2(c) shows E has three friends whereas up to 2 can be monitored by E such that C and F with lowest in-degree are selected. In Fig. 2(c), only two vertices D, G and an edge dg are left in graph, finally we remove them from graph and all MCRs are now 0, as shown in Fig. 2(d). The process is complete and the resultant monitoring points are (B, E, G).

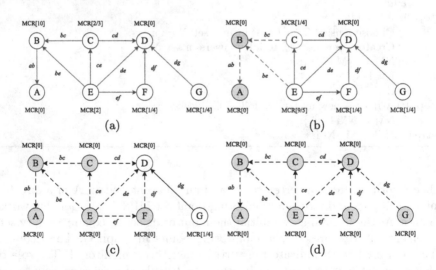

Fig. 2. Illustration of the steps of the MMCRD algorithm. (Color figure online)

5 Experiment and Evaluation

5.1 Dataset Description

In order to analyze how to deploy monitoring points on social network effectively and evaluate MMCRD algorithm, we conduct experiments on datasets derived from Facebook, Twitter, Weibo and synthetic BA scale free networks. On Facebook, Twitter and Weibo, we sample about 2000 users and their relationships relatively as research subjects. We collect all the relationships between these users. In Facebook, the number of relationships is 127841. In Twitter, the number of relationships is 182561 and 13140 relationships in Weibo. In term of synthetic network, we generate scale-free networks based on Barabasi Model [20]. We use Gephi [21], an open-source graph-visualization software to visualize the relationships network of Facebook, Twitter and Weibo respectively as shown in Fig. 3. The detailed topological characteristic of these networks as provided in Table 1.

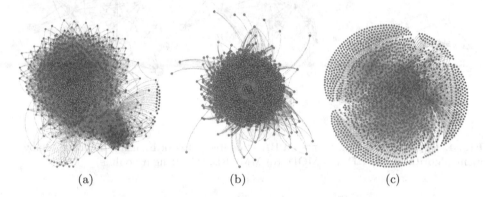

(a)   (b)   (c)

Fig. 3. The relationships network. (a) Facebook relationships network. (b) Twitter relationships network. (c) Weibo relationships network.

Table 1. Topology characteristic of Facebook, Twitter, Weibo Relationships Network.

Topology characteristic	Facebook	Twitter	Weibo
Average degree	116.547	182.65	13.37
Diameter	8	5	8
Density	0.027	0.046	0.003
Modularity	0.25	0.072	0.34
Clustering coefficient	0.195	0.512	0.15
Average path length	1.977	2.088	2.857

5.2 Experiments

Based on these networks, we verify the effectiveness of MMCRD. In order to verify the effectiveness of MMCRD used in monitoring point deployment, we compare MMCRD with RD and MOD. Figure 4 shows the deployment results of RD, MOD and MMCRD in the same BA scale-free network with 20 vertices. In this case, each vertex is allowed to monitor 3 vertices at most. The vertices colored in red represent the monitoring points which are originally included in graph. The vertices colored in blue represent new created monitoring points. From the figure we can see that RD requires 9 monitors, MOD requires 8 monitors and MMCRD requires 7 monitors.

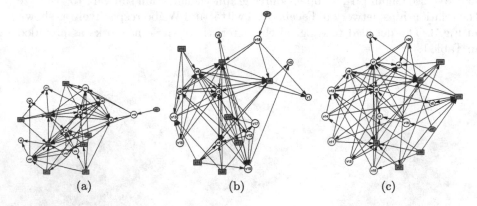

(a)   (b)   (c)

Fig. 4. Deploy monitoring points on BA scale-free network by using different deployment algorithms. (a) RD. (b) MOD. (c) MMCRD. (Color figure online)

Furthermore, we deploy monitoring points on Facebook, Twitter and Weibo and obtain the relationship between monitor capacity and required number of monitors respectively, as shown in Fig. 5. Compared to RD and MOD, given the same timeliness constraint (monitor capacity), MMCRD requires fewer monitors. We only require at most 12% of all users in order to achieve full coverage, the conclusion is coherent in Facebook, Twitter and Weibo three different relationship networks, thus we verify MMCRD is an effective approximation deployment algorithm to achieve full coverage and guarantee timeliness. For example, when the monitor capacity of each user is 50, the required number of monitors of Facebook and Weibo are both about 5% of all users and about 2.5% of all users for Twitter by using MMCRD, therefore using MMCRD to deploy monitoring points can sharply decrease the API calls in data acquisition. Furthermore, if the capacity of each monitor increase to 100, we only acquire 1% monitors in Twitter.

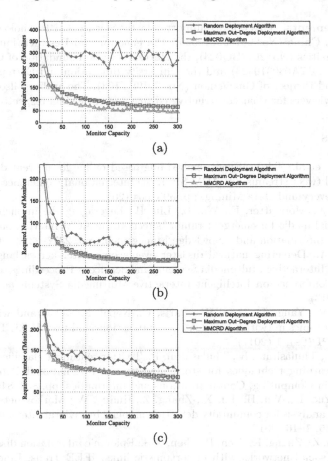

Fig. 5. The relationship between monitor capacity and the required number of monitors. (a) Facebook. (b) Twitter. (c) Weibo.

6   Conclusion and Future Work

Collecting complete and real-time social network data is the key to grasp social public opinion and mine event hot-spots. Smart deployment of monitoring points on social network is crucial in data acquisition. This paper proposed the MMCRD algorithm to deploy monitoring points on social networks based on their MCRs which is depend on out-degree and excess average in-degree of each vertex. Conducting experiment on datasets derived from real-world social networking sites and synthetic social network, we verify the effectiveness of the proposed algorithm which only require to monitor 12% of all users can achieve full coverage and guarantee timeliness. We will further consider the behaviors of different users in the network to improve our algorithm and use a more fine-grained algorithm on various social networks with different user attributes.

Acknowledgements. This work is supported by the Science and Technology Program of Guangzhou, China (No. 201802010025), the Fundamental Research Funds for the Central Universities (No. 2017BQ024), the Natural Science Foundation of Guangdong Province (No. 2017A030310428) and the University Innovation and Entrepreneurship Education Fund Project of Guangzhou (No. 2019PT103). The authors also thank the editors and reviewers for their constructive editing and reviewing, respectively.

References

1. Zhang, C., et al.: TrioVecEvent: embedding-based online local event detection in geo-tagged tweet streams. In: ACM SIGKDD International Conference on Knowledge Discovery and Data Mining, pp. 595–604 (2017)
2. Sampson, J., Morstatter, F., Wu, L., Liu, H.: Leveraging the implicit structure within social media for emergent rumor detection. In: ACM International on Conference on Information and Knowledge Management, pp. 2377–2382 (2016)
3. Zielinski, A.: Detecting natural disaster events on twitter across languages. In: Intelligent Interactive Multimedia Systems and Services - Proceedings of the International Conference on Intelligent Interactive Multimedia Systems and Services, IIMSS (2013)
4. Nugroho, R., Yang, J., Zhao, W., Paris, C., Nepal, S.: What and with whom? Identifying topics in twitter through both interactions and text. IEEE Trans. Serv. Comput. **PP**(99), 1 (2017)
5. Qasem, M., Thulasiram, R., Thulasiram, P.: Twitter sentiment classification using machine learning techniques for stock markets. In: International Conference on Advances in Computing, Communications and Informatics, pp. 834–840 (2015)
6. Liu, W., Yue, K., Wu, H., Fu, X., Zhang, Z., Huang, W.: Markov-network based latent link analysis for community detection in social behavioral interactions. Appl. Intell. **5915**, 1–16 (2017)
7. Xu, Q., Su, Z., Zhang, K., Ren, P., Shen, X.S.: Epidemic information dissemination in mobile social networks with opportunistic links. IEEE Trans. Emerg. Topics Comput. **3**(3), 399–409 (2015)
8. Pavlopoulou, M.E.G., Tzortzis, G., Vogiatzis, D., Paliouras, G.: Predicting the evolution of communities in social networks using structural and temporal features. In: International Workshop on Semantic and Social Media Adaptation and Personalization, pp. 1–6 (2017)
9. Rife, S.C., Cate, K.L., Kosinski, M., Stillwell, D.: Participant recruitment and data collection through facebook: the role of personality factors. Int. J. Soc. Res. Methodol. **19**(1), 69–83 (2016)
10. Buzzi, M.C., et al.: Facebook: a new tool for collecting health data? Multimedia Tools Appl. **76**, 1–24 (2016)
11. Nia, M.H.: Social big data analytics of consumer choices: a two sided online platform perspective. Social Science Electronic Publishing (2017)
12. Carbonneau, R., Vahidov, R.: A multi-attribute bidding strategy for a single-attribute auction marketplace. Expert Syst. Eith Appl. **43**(C), 42–50 (2016)
13. Yasmina, D., Hajar, M., Hassan, A.M.: Using Youtube comments for text-based emotion recognition. Procedia Comput. Sci. **83**, 292–299 (2016)
14. Malik, H., Tian, Z.: A framework for collecting youtube meta-data. Procedia Comput. Sci. **113**, 194–201 (2017)
15. Cao, Y., Long, M., Wang, J., Yu, P.S.: Correlation hashing network for efficient cross-modal retrieval (2016)

16. Singla, A., Tschiatschek, S., Krause, A.: Noisy submodular maximization via adaptive sampling with applications to crowdsourced image collection summarization. Computer Science (2016)
17. Steiger, E., Albuquerque, J.P., Zipf, A.: An advanced systematic literature review on spatiotemporal analyses of Twitter data. Trans. GIS **19**(6), 809–834 (2016)
18. Xiao, X., Attanasio, A., Chiusano, S., Cerquitelli, T.: Twitter data laid almost bare: an insightful exploratory analyser. Expert Syst. with Appl. **90**(2017), 09 (2017)
19. Jendryke, M., Balz, T., Liao, M.: Big location based social media messages from China's Sina Weibo network: collection, storage, visualization, and potential ways of analysis. Trans. GIS **21**(4), 825–834 (2017)
20. Barabsi, A.: Scale-free networks: a decade and beyond. Science **325**(5939), 412–413 (2009)
21. Bastian, M., Heymann, S., Jacomy, M.: Gephi: an open source software for exploring and manipulating networks. ICWSM **8**(2009), 361–362 (2009)

Deep Learning for Knowledge-Driven Ontology Stream Prediction

Shumin Deng[1], Jeff Z. Pan[2], Jiaoyan Chen[3], and Huajun Chen[1(✉)]

[1] College of Computer Science and Technology, Zhejiang University,
Hangzhou, China
{231sm,huajunsir}@zju.edu.cn
[2] Department of Computing Science, The University of Aberdeen, Aberdeen, UK
jeff.z.pan@abdn.ac.uk
[3] Department of Computer Science, University of Oxford, Oxford, UK
jiaoyan.chen@cs.ox.ac.uk

Abstract. Time series prediction with data stream has been widely studied. Current deep learning methods *e.g.*, Long Short-Term Memory (LSTM) perform well in learning feature representations from raw data. However, most of these models can narrowly learn semantic information behind the data. In this paper, we revisit LSTM from the perspective of Semantic Web, where streaming data are represented as ontology sequences. We propose a novel semantic-based neural network (STBNet) that (i) enriches the semantics of data stream with external text, and (ii) exploits the underlying semantics with background knowledge for time series prediction. Previous models mainly rely on numerical representation of values in raw data, while the proposed STBNet model creatively integrates semantic embedding into a hybrid neural network. We develop a new attention mechanism based on similarity among semantic embedding of ontology stream, and then we combine ontology stream and numerical analysis in the deep learning model. Furthermore, we also enrich ontology stream in STBNet, where Convolutional Neural Networks (CNNs) are incorporated in learning lexical representations of words in the text. The experiments show that STBNet outperforms state-of-the-art methods on stock price prediction.

Keywords: Ontology stream · Deep learning · Time series prediction

1 Introduction

Data stream learning has been widely studied for extracting knowledge from evolving data. Time series prediction is an important component of data stream learning. It has been widely applied in forecasting stock price [16], indoor temperature [15], etc. In most applications of time series prediction, target output results depend on exogenous input terms (driving series). What's more, these driving series are nonlinear in most cases. To address this issue, deep learning methods have been proposed accordingly in recent years, which can learn

© Springer Nature Singapore Pte Ltd. 2019
J. Zhao et al. (Eds.): CCKS 2018, CCIS 957, pp. 52–64, 2019.
https://doi.org/10.1007/978-981-13-3146-6_5

the nonlinear relationship automatically. Recurrent neural networks (RNNs) [8,18,19], a type of deep neural network specially designed for sequence modeling, are powerful in discovering the dependency in sequence data. Traditional RNNs, however, suffer from the problem of vanishing gradients and thus have difficulty in capturing long-term dependencies. The Long Short-Term Memory (LSTM) RNN [9] has overcome this limitation and works well on sequence data with long-term dependencies. There have already been some works on learning feature representation from sequence data, with the use of LSTM methods. *e.g.* [15] has proposed TreNet, a novel end-to-end hybrid neural network, to learn local and global contextual features for predicting the trend of time series.

Despite substantial efforts have been made for time series prediction, most of these approaches only learn the data feature representation of time series, *e.g.*, the nonlinear relationship among driving series, while ignore the underlying semantic impacts. In fact, semantical information can be regarded as prior knowledge in data stream learning. For example, Apple Inc. and eBay Inc. both belong to "Information Technology" Sector and "Internet Software & Services" Sub Industry, and their headquarters are both located in California. If an IT-related event occurs, stock price of these two companies may have similar fluctuation accordingly. Essentially, the key is to availably embed background knowledge into training models. In the semantic Web, data is interpreted in ontologies and its ordered sequence is represented as an ontology stream [17], which is proved to be an effective method for semantic embedding. [13,14] tackle predictive reasoning as a correlation and interpretation of past semantic-augmented data over exogenous ontology streams. [7] exploit the semantics of such streams to tackle the problem of concept drift.

Apart from discovering knowledge through predictive reasoning, extracting semantic features from corresponding text also makes sense in time series prediction. The effectiveness can be improved if there is latent professional expertise in text which is understood accurately. Some works also have been done on semantic feature extraction from text. *e.g.* [6] introduce Sem-CNN, a wide and deep CNN model that can leverage concepts contained in crisis-related social media content. Besides, [21] propose a novel deep architecture to utilize both structural and textual information of entities, which can enrich semantic information of entities, in contrast to most works focused on symbolic representation of knowledge graph with structure information.

Inspired by traditional data feature learning with LSTM methods, semantic reasoning with ontology stream, and semantic feature learning with text, we propose a novel neural network with semantic embedding (STBNet). As traditional methods almost ignore semantics of background knowledge, it is necessary to demonstrate its impacts on time series prediction. Thus, on the basis of LSTM framework, we encode semantics of background knowledge in models. Furthermore, we train a CNN model to exploit feature representation from text stream, whose output serves as an enrichment factor for semantic embedding.

The next section reviews preliminary of our work, including ontology stream learning problems and text stream learning, as well as data stream learning with deep learning models. Section 3 briefly formulates our problem and gives the model overview. Section 4 presents semantic embedding and prediction for ontology stream w.r.t text embedding. Section 5 presents the experiments and evaluation. Section 6 makes a conclusion of the paper and discusses the future work.

2   Related Work

Time series is a sequence of data in time order. The semantics of data are represented using an ontology, and can be enriched with text. In this section, we briefly introduce (i) ontology stream learning problems, (ii) text stream learning, and (iii) traditional deep learning models for time series prediction.

2.1   Ontology Stream Learning Problems

A Description Logics ($\mathcal{DL}$) [2] ontology $\mathcal{O} \doteq \langle \mathcal{T}, \mathcal{A} \rangle$ is composed of TBox $\mathcal{T}$ and ABox $\mathcal{A}$. We represent evolving knowledge by ontologies with different versions [10]. Besides, data (ABox $\mathcal{A}$) evolves over time while the schema (TBox $\mathcal{T}$) remains unchanged. We define the changing ontology sequence as the ontology stream [17].

Definition 1 *(Ontology Stream)*
An ontology stream $\mathcal{O}_m^n$ from timestamp m to timestamp n is a sequence of ontologies $(\mathcal{O}_m^n(m), \mathcal{O}_m^n(m+1), \cdots, \mathcal{O}_m^n(n))$, where $m, n \in N$ and $m < n$.

$\mathcal{O}_m^n(i)$ is a snapshot of an ontology stream $\mathcal{O}_m^n$ at timestamp i, referring to a set of axioms in a $\mathcal{DL}$. Therefore, a transition from $\mathcal{O}_m^n(i)$ to $\mathcal{O}_m^n(i+1)$ is regarded as an ABox update. We consider that the $\mathcal{O}_0^n$ is composed of all snapshots in time order. In order to extract more knowledge from $\mathcal{O}_0^n$, $\mathcal{O}_m^n[i, j]$ is denoted as a windowed stream of $\mathcal{O}_m^n$ between timestamp i and j with $i \leq j$.

2.2   Text Stream Learning

As expressive knowledge in ontology stream are often incomplete, it is helpful to enrich the knowledge with other sources. [6] introduce a wide and deep CNN model that can leverage concepts contained in text. [21] propose a novel deep architecture to utilize both structural and textual information of entities, which can enrich semantic information of entities. Inspired by their work, we enrich semantic information of entities by leveraging concepts contained in text. In our model, text is represented in the form of stream, corresponding to ontology stream.

Definition 2 *(Text Stream)*
An text stream $\mathcal{X}_m^n$ is a sequence of text corresponding to ontology stream $\mathcal{O}_m^n$. $\mathcal{X}_m^n$ is a text-set $(\mathcal{X}_m^n(m), \mathcal{X}_m^n(m+1), \cdots, \mathcal{X}_m^n(n))$ from timestamp m to timestamp n, where $m, n \in N$ and $m < n$.

2.3 Time Series Forecasting Models

There are various models for time series prediction, and the Long Short-Term Memory (LSTM) [9] model has performed state-of-the-art results among them. TreNet [15] is a novel end-to-end hybrid neural network. The model learns the dependency in the historical trend sequence with the use of LSTM. Besides, as the recent raw data points of time series can affect the evolving of trend, TreNet learns local features from raw data of time series with CNN. Finally, the feature fusion layer combines the output representations from LSTM and CNN. TreNet has shown that LSTM can be attached to external memory. Through this discovery, we get the idea that background knowledge can be embedded, and stored in external memory.

[16] have proposed a dual-stage attention based recurrent neural network (DA-RNN), which is composed of two stages. In the first stage, the model develops a new attention mechanism to adaptively extract the relevant driving series at each timestamp. In the second stage, a temporal attention mechanism is used to select relevant encoder hidden states across all timestamps. These two attention models are well integrated within a LSTM network and can be jointly trained. DA-RNN inspires us that the attention mechanism can be applied to relevant driving series. In DA-RNN, all the attention distribution is trained only with input of data representation, while ignores the underlying relations among driving series. It also enlightens us that attention mechanism can be improved by embedding background knowledge into training process.

3 Problem Formulation and Model Overview

In this section, we introduce the formal definition used in this work and the problem we aim to study. Then we present the overview of the proposed STBNet.

3.1 Problem Formulation

We define n driving series as n sequences of data points, denoted by $s = \{s_0, s_1, \cdots, s_{n-1}\}$, where each sequence s_k represents real-time prices of a stock with time span T. Analogously, the target series is denoted by $p = \{p_0, p_1, \cdots, p_{T-1}\}$, where the subscript t represents the timestamp.

Meanwhile, as we introduced in Sect. 2, time series data are represented as the ontology stream. $\mathcal{P}$, $\mathcal{S}$, and $\mathcal{X}$ denote the ontology stream of the S&P 500 index, the stock price, and text data respectively. Through ontology stream learning, we generate the entailment vector, denoted by $e = \{e_0, e_1, \cdots, e_m\}$. Each element in e indicates a factor acts on the target series, and m is the factor quantity.

Besides, we also import background knowledge, which are in the form of triples. These triples are embedded as a set of vectors. The vector of the target entity is defined as v_{sp}, and that of driving entities is denoted by $v = \{v_0, v_1, \cdots, v_{n-1}\}$. d represents the semantic distances among the target entity and driving entities, and the attention vector α can be calculated with d, denoted by $\alpha = \{\alpha_0, \alpha_0, \cdots, \alpha_{n-1}\}$.

Then, we aim to forecast the target series p by learning a function $p = F(s, e, \alpha)$. In this paper, we propose a neural network based on semantic embedding, which can make prediction on target values in consideration of semantics.

3.2 Overview of STBNet

We go on to give an overview of the model architecture, which is shown in Fig. 1.

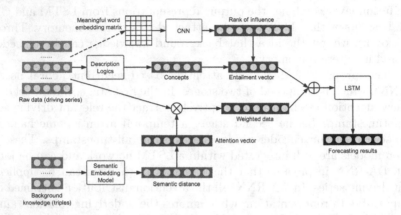

Fig. 1. Illustration of the STBNet model architecture (in a snapshot of ontology stream)

The original input of STBNet includes raw data and background knowledge, as well as corresponding text data. As for raw data, we not only utilize numerical representation of values, but also extract high level concepts and entailments generated from data features. The semantic embedding of raw data will be introduced in details, cf. Sect. 4.2. In order to enrich semantic information of entailments, text data are fed into a CNN model for lexical feature extraction (cf. Sect. 4.3 for more details). When it comes to background knowledge, we have calculated the semantic distances among entities, through ontology design and semantic embedding. And then, we have built an attention mechanism with the semantic distances. Raw data should be multiplied by the attention vector before fed into a LSTM model. Then, the model input is updated, including the lexical feature vectors, the bag of entailments, and the attention-based data through semantic embedding, and all the updated input will be fed into a LSTM model, to get forecasting results as output.

4 Method

Our proposed method mainly includes three parts: (i) embedding the background knowledge of the ontology stream with entailment vectors, (ii) enriching background knowledge with text embedding, and (iii) forecasting the target object based on semantic embedding and numerical analysis.

4.1   Semantic Embedding

In this paper, the semantics of ontology stream contains two kinds of knowledge: (i) entailments of ontology stream, encoded by the entailment vector (cf. Definition 3), and (ii) semantic attention to different background knowledge, encoded by attention vector (cf. Definition 4).

Definition 3 (Entailment Vector)
Assumed that $\mathcal{G} \doteq \{g_1, g_2, \cdots, g_m\}$ is all Abox entailments in an ontology stream $\mathcal{P}_0^n$, and $P_0^n(i)$ is a snapshot in P_0^n. The entailment vector of a snapshot $P_0^n(i)$ is a vector with dimension m, denoted by e_i. e_i is composed of e_{ij}, $\forall j \in [0, m]$.

$$e_{ij} = \begin{cases} 1 & if \quad \mathcal{T} \cup \mathcal{A} \cup P_0^n(i) \models g_j \\ 0 & otherwise \end{cases} \tag{1}$$

where g_j is one of Abox entailments.

The entailment bags can be regarded as different factors which can have significant effects on the ontology stream. For ontology stream $\mathcal{P}_0^n$ of S&P 500 index, the bags of entailments are composed of (i) sector, (ii) sub industry, (iii) city, and (iv) state, associated with the stock and its relevant security.

Definition 4 (Attention Vector)
Given the entity vector $\boldsymbol{v}_k$ for each individual of driving series, as well as $\boldsymbol{v}_{sp}$ for the stock index, then the attention vector of driving series is calculated by Eq. (2). (The entity vectors can be trained from an embedding model (e.g., TransE [5]).)

$$\alpha_k = \frac{exp(\lceil max(\boldsymbol{d}) - |\boldsymbol{v}_{sp} - \boldsymbol{v}_k|\rceil)}{\sum_{i=1}^{q} exp(\lceil max(\boldsymbol{d}) - |\boldsymbol{v}_{sp} - \boldsymbol{v}_i|\rceil)} \tag{2}$$

where $|\cdot|$ represents a vector's Euclidean distance, $|\boldsymbol{v}_{sp} - \boldsymbol{v}_k|$ is the semantic distance between the stock index and the stock individual (and $max(\boldsymbol{d})$ is the maximum among them), $\lceil\cdot\rceil$ denotes a ceil function, $k \in [0, q-1]$, and q is the quantity of driving series individuals. A softmax function is applied here, in order to ensure the sum of α_k equals to 1. In Eq. (2), the less $|\boldsymbol{v}_{sp} - \boldsymbol{v}_k|$ is, the larger α_k will be. The constructing process of the attention vector is shown in Fig. 2.

4.2   Text Embedding for Ontology Stream

Though ontology stream learning performs well in utilizing background knowledge, the performance improvement is still limited due to the lack of knowledge source. In this situation, we introduce our method of text embedding, which incorporates text as additional knowledge for neural network training.

Sentence Extraction from Text Stream. Text data are usually poorly structured, making it difficult to extract semantic information precisely. Thus, compared to embed a whole sentence, to handle with meaningful words in it is much feasible. We extract named-entities in the text and map them to their associated semantic concepts, with the use of multiple semantic knowledge bases including DBpedia [1] and Freebase [4]. We extract the sentence from text stream, if words in it have associated semantic concepts in knowledge bases.

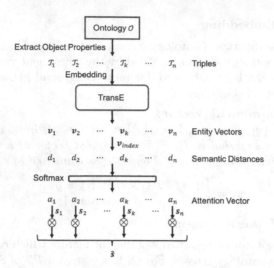

Fig. 2. Constructing attention vector for each stock individual.

Text Stream Learning with a CNN Model. The steps of text stream learning are divided into (i) word vectorization, (ii) text feature extraction, and (iii) results acting on entailment vectors (cf. Definition 3).

In the step of word vectorization, we split the sentences into a set of meaningful words. As the text stream contains real-time text information, the word-sets are also generated by time. In this paper, the time span is daily effective time in the stock market. Each meaningful word in a sentence is mapped into a vector, and the word embedding matrix is generated by concatenating these word vectors. Next, in the step of text feature extraction, the temporal word embedding matrix will be fed into a CNN model. The architecture of CNN is presented in Fig. 3.

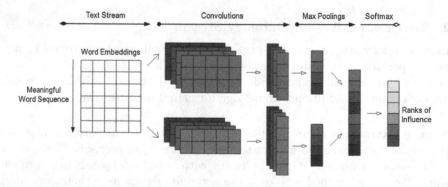

Fig. 3. The Convolutional Neural Network (CNN) for text features extraction

As shown in Fig. 3, the CNN model is divided into 4 layers, including the input layer, the convolution layer, max pooling layer and a softmax layer. The input data are the word embedding matrices generated from text stream. The length of a matrix is the size of a word vector, and the width is quantity of words extracted from daily text stream. As the word quantity may be various in different days, minimum is chosen. In the convolutional layer, there are multiple convolutional filters of varying sizes to generate feature vectors. Max pooling is performed on each layer and a softmax classifier is used for rank evaluation.

Finally, the feature vectors learned from the CNN serve as a generation factor for bags of entailments. The output of CNN is a set of numbers in $[-2.5, 2]$ with interval 0.1, which reflects the fluctuating trend of the S&P 500 index values. Negative values indicate the index value is likely to decrease, positive values show an increase trend, and value size demonstrates the possibility scale.

4.3 Semantic Prediction

We have selected LSTM as our main model, for its great power in discovering long-term dependencies in sequence data. In our proposed STBNet model, the input data includes: (i) attention-based [3] input of stock prices and (ii) entailment vectors w.r.t text stream and concepts extracted in raw data.

With the attention vectors calculated by Eq. (2), we have extracted the driving series through

$$\tilde{s} = [\alpha_0 s_0, \alpha_1 s_1, \cdots, \alpha_k s_k, \cdots, \alpha_n s_n] \tag{3}$$

where s_k is one of the driving series individuals, α_k is the corresponding weight value, $k \in \{0, 1, \cdots, n-1\}$, and n is the quantity of driving series individuals.

We have concatenated entailment vector e to $\tilde{s}$, and the updated input of LSTM is $[\tilde{s}, e]$. During the training stage, $s' = [\tilde{s}, e]$ is fed into LSTM, and the result for target forecasting is denoted by $\hat{p} = F(s')$.

Then, we adopt the squared error function plus a regularization term as the cost function, denoted by

$$\mathcal{L} = \frac{1}{T} \sum_{k=1}^{T} (p_k - \hat{p}_k)^2 + \lambda \|W\|_2 \tag{4}$$

where W represent all the weight parameters in STBNet, and λ is a hyperparameter for the regularization term.

5 Experiments

5.1 Dataset and Setup

Stock Price Data. The stock price data[1] are extracted from the Google Finance API. It contains minutely price records of 500 stocks from S&P 500,

---

[1] http://files.statworx.com/sp500.zip.

ranging from 3rd April to 31th August in 2017. Note that they are cleaned for bank holidays, and aligned with Twitter text by time. In order to align the stock price data with real-time text data, we have chosen the time span of one day.

Background Knowledge of Stocks. It contains correlative companies information of S&P 500 component stocks, which is stored in a knowledge graph. Each stock is issued by corresponding *Security*, which has *Sector*, *Sub Industry*, and *Address of Headquarters*. These background knowledge are all extracted from Wikipedia [20].

Real-Time Text Data for Stocks. It contains social media data related to S&P 500 index and stocks when the stock market opening. We choose tweet data for its wide source and popularity. Sentences in tweets can be dirty and incomplete, noisy and poorly structured. Therefore, we have cleared and formalized the tweets data and aligned it with raw value data by time. We have extracted 98617 tweets with respect to S&P 500 index, from 3rd, April, 2017 to 31st, August, 2017. As the meaningful words in one minute or one hour are a bit limited, we choose to do text stream learning on daily tweets.

5.2   Training and Evaluation

In training process, stochastic gradient descent (SGD) [11] together with Momentum Optimizer are used to train the STBNet model. We train the CNN model [12] of STBNet with 706 meaningful word embeddings vectors, in use of 128 convolution filters of sizes $[2, 3, 4, 5]$. In LSTM model of STBNet, We use a dropout rate of 0.5 to avoid overfitting and the learning rate is 0.0001. We split the raw value dataset into train set, valid set and test set respectively, with ratio 0.8, 0.1 and 0.1.

In order to validate the effectiveness of importing background knowledge and text stream in time series prediction, we evaluate the predictive performance of STBNet and baseline models in terms of three different evaluation metrics, including (i) mean absolute error (MAE), (ii) mean absolute percentage error (MAPE), and (iii) root mean squared error (RMSE).

5.3   Experiment Results

The baseline models are general models for time series prediction, including ARIMA and LSTM. We also compare them with our proposed methods: SE-LSTM and TE-LSTM. Finally, these four models are all compared with STBNet.

(i) ARIMA (autoregressive integrated moving average) is a machine learning model. The basic idea of ARIMA is that the data sequence formed by the predicted object over time is considered as a random sequence, and a certain mathematical model is used to approximately describe the sequence. As the model is suitable for stationary series, non-stationary S&P 500 index data are transformed into stationary one before regression.

(ii) LSTM, a RNN model, is one of the mainstream architectures for time series prediction. LSTM does well in discovering dependencies in sequence data, especially data with long-term dependencies.

(iii) SE-LSTM is inherited from a LSTM model, and can be regarded as a primary version of our proposed STBNet. Compared with traditional LSTM models, SE-LSTM imports an attention mechanism on input driving series, based on semantic embeddings of background knowledge.

(iv) TE-LSTM is also inherited from a LSTM model, and can be regarded as another primary version of our proposed STBNet. Compared to traditional LSTM models, TE-LSTM enriches its input driving series with entailment vectors w.r.t text embedding.

Our proposed STBNet model is a combination of SE-LSTM and TE-LSTM, which enriches the input driving series with the attention vector and entailment vectors.

The effectiveness of STBNet and baseline methods over the S&P 500 dataset are shown in Table 1.

Table 1. Time series prediction results over the S&P 500 stock dataset (best performance displayed in **boldface**)

Model		MAE	MAPE ($\times 10^{-2}$)	RMSE
General baseline	ARIMA	17.842	7.352	21.060
	LSTM	3.368	1.806	4.382
Our method	SE-LSTM	2.761	1.417	3.446
	TE-LSTM	2.768	1.416	3.443
	STBNet	**2.758**	**1.411**	**3.254**

In Table 1, we observe that the performance of ARIMA is greatly worse than LSTM based approaches. This is because ARIMA only considers the target series, while ignores the driving series. In order to visualize the difference among the traditional LSTM and our proposed models, we compare some prediction sample results to true values, and the comparison is shown in Figs. 4 and 5.

As can be seen in Fig. 4, both SE-LSTM and TE-LSTM shows better performance than LSTM. It is because SE-LSTM and TE-LSTM are in view of background knowledge and text stream respectively. Such a result validates our hypothesis, which is that semantic knowledge can enhance the effectiveness of time series prediction under certain circumstances.

In Fig. 5, we compare our proposed STBNet with LSTM. STBNet combines SE-LSTM and TE-LSTM, which considers both background knowledge and text stream. As a result, STBNet performs the much better enhance-effect than LSTM. In the stock market, S&P 500 index can fluctuate irregularly, therefore, it can be insufficient to consider time dependency merely. The results show that background knowledge and text stream can jointly improve the enhance-effect.

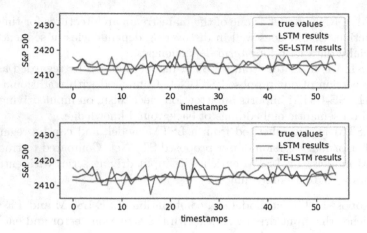

Fig. 4. Comparison of performance with LSTM and SE-LSTM/TE-LSTM. (best viewed in color)

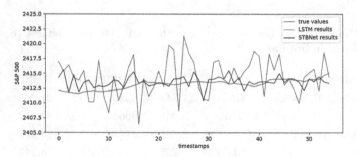

Fig. 5. Comparison of performance with LSTM and STBNet. (best viewed in color)

6   Conclusion and Future Work

In this paper, we propose a novel semantic embedding based neural network (STBNet), to tackle the problem of time series prediction. Traditional methods mostly focus on learning feature representation of raw data, while pay little attention to the underlying semantic impact. In this work, we first interpret time-series data as ontology stream, and then exploit the semantics of such streams in neural network training. We develop a new attention mechanism based on background knowledge embedding. We also enrich the background knowledge with text stream for more accurate prediction. The experiments have shown that STBNet can outperform the state-of-the-art models in stock price prediction. It has validated that semantics input can improve the performance of time series prediction in some deep learning models.

In the future, we will try to replace tweets with other text datasets, for example, news data, in order to select text stream with more enriched semantic knowledge. Besides, we are about to extract structured events from text, and jointly train event tuples and triples in the knowledge base, to get more enriched semantic information from text stream. Furthermore, we will seek for other suitable scenes and do experiments on additional datasets from domains other than finance.

References

1. Auer, S., Bizer, C., Kobilarov, G., Lehmann, J., Cyganiak, R., Ives, Z.: DBpedia: a nucleus for a web of open data. In: Aberer, K., et al. (eds.) ASWC/ISWC -2007. LNCS, vol. 4825, pp. 722–735. Springer, Heidelberg (2007). https://doi.org/10.1007/978-3-540-76298-0_52
2. Baader, F., Brandt, S., Lutz, C.: Pushing the $\mathcal{EL}$ envelope (2005)
3. Bahdanau, D., Cho, K., Bengio, Y.: Neural machine translation by jointly learning to align and translate. Computer Science (2014)
4. Bollacker, K., Evans, C., Paritosh, P., Sturge, T., Taylor, J.: Freebase: a collaboratively created graph database for structuring human knowledge. In: SIGMOD Conference, pp. 1247–1250 (2008)
5. Bordes, A., Usunier, N., Weston, J., Yakhnenko, O.: Translating embeddings for modeling multi-relational data. In: International Conference on Neural Information Processing Systems, pp. 2787–2795 (2013)
6. Burel, G., Saif, H., Alani, H.: Semantic wide and deep learning for detecting crisis-information categories on social media. In: d'Amato, C., et al. (eds.) ISWC 2017. LNCS, vol. 10587, pp. 138–155. Springer, Cham (2017). https://doi.org/10.1007/978-3-319-68288-4_9
7. Chen, J., Lécué, F., Pan, J.Z., Chen, H.: Learning from ontology streams with semantic concept drift. In: IJCAI, pp. 957–963 (2017)
8. Elman, J.L.: Distributed representations, simple recurrent networks, and grammatical structure. Mach. Learn. **7**(2–3), 195–225 (1991)
9. Hochreiter, S., Schmidhuber, J.: Long short-term memory. Neural Comput. **9**(8), 1735–1780 (1997)
10. Huang, Z., Stuckenschmidt, H.: Reasoning with multi-version ontologies. In: The Semantic Web - ISWC 2005, International Semantic Web Conference, ISWC 2005, Galway, Ireland, 6–10 November 2005, Proceedings, pp. 398–412 (2005)
11. Ketkar, N.: Stochastic gradient descent. Optimization (2014)
12. Kim, Y.: Convolutional neural networks for sentence classification. Eprint Arxiv (2014)
13. Lécué, F., Pan, J.Z.: Predicting knowledge in an ontology stream. In: IJCAI, pp. 2662–2669 (2013)
14. Lécué, F., Pan, J.Z.: Consistent knowledge discovery from evolving ontologies. In: AAAI, pp. 189–195 (2015)
15. Lin, T., Guo, T., Aberer, K.: Hybrid neural networks for learning the trend in time series. In: Twenty-Sixth International Joint Conference on Artificial Intelligence, pp. 2273–2279 (2017)
16. Qin, Y., Song, D., Chen, H., Cheng, W., Jiang, G., Cottrell, G.: A dual-stage attention-based recurrent neural network for time series prediction, pp. 2627–2633 (2017)
17. Ren, Y., Pan, J.Z.: Optimising ontology stream reasoning with truth maintenance system. In: ACM International Conference on Information and Knowledge Management, pp. 831–836 (2011)

18. Rumelhart, D.E., Hinton, G.E., Williams, R.J.: Learning representations by back-propagating errors. Nature **323**(6088), 533–536 (1986)
19. Werbos, P.: Backpropagation through time: what it does and how to do it. Proc. IEEE **78**(10), 1550–1560 (1990)
20. Milne, D., Witten, I.H.: Learning to link with wikipedia. In: ACM Conference on Information and Knowledge Management, pp. 509–518 (2008)
21. Xu, J., Chen, K., Qiu, X., Huang, X.: Knowledge graph representation with jointly structural and textual encoding, pp. 1318–1324 (2016)

DSKG: A Deep Sequential Model for Knowledge Graph Completion

Lingbing Guo[1,2], Qingheng Zhang[1], Weiyi Ge[2], Wei Hu[1(✉)], and Yuzhong Qu[1]

[1] State Key Laboratory for Novel Software Technology,
Nanjing University, Nanjing, China
lbguo.nju@gmail.com, qhzhang.nju@gmail.com, {whu,yzqu}@nju.edu.cn
[2] Science and Technology on Information Systems Engineering Lab, Nanjing, China
geweiyi@163.com

Abstract. Knowledge graph (KG) completion aims to fill the missing facts in a KG, where a fact is represented as a triple in the form of $(subject, relation, object)$. Current KG completion models compel two-thirds of a triple provided (e.g., *subject* and *relation*) to predict the remaining one. In this paper, we propose a new model, which uses a KG-specific multi-layer recurrent neutral network (RNN) to model triples in a KG as sequences. It outperformed several state-of-the-art KG completion models on the conventional entity prediction task for many evaluation metrics, based on two benchmark datasets and a more difficult dataset. Furthermore, our model is enabled by the sequential characteristic and thus capable of predicting the whole triples only given one entity. Our experiments demonstrated that our model achieved promising performance on this new triple prediction task.

Keywords: Knowledge graph completion · Deep Sequential model
Recurrent neutral network

1 Introduction

Knowledge graphs (KGs), such as Freebase [2] and WordNet [12], typically use triples, in the form of $(subject, relation, object)$ (abbr. (s, r, o)), to record billions of real-world facts, where s, o denote entities and r denotes a relation between s and o. Since current KGs are still far from complete, the KG completion task makes sense. Previous models focus on a general task called entity prediction (a.k.a. link prediction) [3], which asks for completing a triple in a KG by predicting o (or s) given $(s, r, ?)$ (or $(?, r, o)$). Figure 1a shows an abstract model for entity prediction. Input s, r are firstly projected by some vectors or matrices, and then combined to a continuous representation v_o to predict o.

Although previous models perform well on entity prediction, they may still be inadequate to complete a KG. Let us assume that a model can effectively complete an entity s given a relation r explicitly. If we do not provide any relations, this model is incompetent to fill s, because it is incapable of choosing

© Springer Nature Singapore Pte Ltd. 2019
J. Zhao et al. (Eds.): CCKS 2018, CCIS 957, pp. 65–77, 2019.
https://doi.org/10.1007/978-981-13-3146-6_6

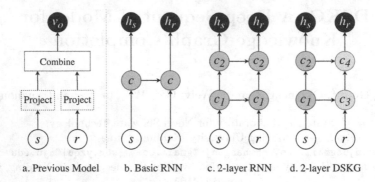

a. Previous Model   b. Basic RNN   c. 2-layer RNN   d. 2-layer DSKG

Fig. 1. Different models for entity prediction. White and black circles denote input and output vectors, respectively. c denotes an RNN cell and h denotes a hidden state. DSKG uses c_1, c_2 to process entity s, and c_3, c_4 to process relation r. All of them are different RNN cells. (Color figure online)

which relation to complete this entity. Actually, the underlying data model of KGs does not allow the existence of any incomplete tuple (s, r).

The recurrent neural network (RNN) is a neural sequence model, which has achieved good performance on many natural language processing (NLP) tasks, such as language modeling and machine translation [7,16]. Triples in a KG can be approximately regarded as simple sentences of length 3. For example, a triple (*USA, contains, NewYorkCity*) can be transformed to a sentence *"USA contains New York City"*. This enlightens us to use RNNs to model KGs. However, we are still challenged by the following problems: (i) Triples are not natural language. They model the complex structure with a fixed expression (s, r, o). Such short sequences may be insufficient to provide enough context for prediction. Meanwhile, it is costly and difficult to construct valuable long sequences from massive paths; (ii) Relations and entities are two different types of elements that appear in triples in a fixed order. It is inappropriate to treat them as the same type.

To solve the aforementioned problems, we propose DSKG (Deep Sequential model for KG), which employs RNNs with a new structure and uses a KG-specific sampling method for training. We design a basic RNN model as the initial version to illustrate our idea (see Fig. 1b). This basic model takes input s, r as the same type of elements and recurrently processes them. An RNN cell is denoted by c, which receives its previous hidden state and the current element as input to predict the next. The cell in the entity layer processes entities like s, while the cell in the relation layer processes relations like r. In this model, there only exists one cell to sequentially process all input elements, so s, r are fed to the same cell c to obtain their respective output. We then use h_s to predict the relations of s and h_r to predict the objects of $s \rightarrow r$.

The basic RNN model may be not model complex structures well, because it only uses a single RNN cell to process all input sequences. In the NLP area, researchers usually stack multiple RNN cells to improve the performance. Here, we borrow this idea to build a multi-layer RNN model (see Fig. 1c). But still, this model cannot precisely model triples in a KG, since entities and relations have very different characteristics.

As depicted in Fig. 1d, the proposed DSKG employs respective multi-layer RNNs to process entities and relations. Specifically, DSKG uses independent RNN cells for the entity layer and the relation layer, i.e., c_1, c_2, c_3, c_4 in the figure are all different RNN cells. We believe that this KG-specific architecture can achieve better performance when relations are diverse and complex. Because DSKG considers predicting entities (or relations) as a classification task, we also propose a customized sampling method that samples negative labels according to the type of current training label. Furthermore, DSKG has the ability to predict the relations of one entity, which inspires us to employ a method to enhance the results of entity prediction by relation prediction. For example, when predicting (*USA, contains, ?*), the model can automatically filter entities like people or movies, because these entities are not related to relation *contains*.

We conducted the entity prediction experiment on two benchmark datasets. The results showed that DSKG outperformed several state-of-the-art models in terms of many evaluation metrics. Furthermore, we evaluated DSKG on FB15K-237 [14]. The results demonstrated that DSKG outperformed the other models. Additionally, we designed a new KG completion experiment for triple prediction, as a complement to entity prediction. We demonstrated that, as compared with the general multi-layer RNN models, DSKG also achieved superior results. Our source code, datasets and experimental results are available online[1].

2 Related Work

2.1 TransE-Like Models

TransE [3] represents entities and relations as k-dimensional vectors in a unified space, and models a triple (s, r, o) as $s + r \approx o$. TransE works well for the one-to-one relationship, but fails to model more complex (e.g., one-to-many) relationships. TransH [17] tries to solve this problem by regarding each relation r as a vector on a hyperplane whose normalization vector is w_r. It projects entities s, o to this hyperplane by w_r. TransR [11] uses relation-specific matrices to project entities. For each relation r, it creates a matrix W_r to project s, o by W_r. TransR also adopts the same energy function.

PTransE [10] leverages additional path information for training. For example, if there exist two triples $(e_1, r_1, e_2), (e_2, r_2, e_3)$, which can be regarded as a path in a KG, and another triple (e_1, r_x, e_3) holds simultaneously, then $e_1 \rightarrow r_1 \rightarrow e_2 \rightarrow r_2 \rightarrow e_3$ is a valuable path and recorded as (e_1, r_1, r_2, e_3). However, preparing desirable paths requires to iterate over all possible paths, thus this process may consume enormous resources for large KGs. Consequently, PTransE and other path-based models may be inefficient to model large KGs.

All the aforementioned models choose to minimize an energy function that is used in or similar to TransE. Moreover, TransR and PTransE use pre-trained entity and relation vectors from TransE as initial input.

---

[1] https://github.com/nju-websoft/DSKG.

2.2 Other Models

Some models are different from the TransE-like models. DISTMULT [20] is as simple as TransE, but uses a completely different energy function. It is based on a bilinear model [13], and represents each relation as a diagonal matrix. ComplEx [15] extends DISTMULT with the complex embedding technique.

Node+LinkFeat (in short, NLFeat) [14] can also be regarded as a path-based model similar to PTransE, but only needs to extract paths of length 1 for constructing node and link features. Although it uses paths of length 1, it still consumes considerable resources for large KGs. NeuralLP [21] aims at learning probabilistic first-order logical rules for KG reasoning. For each relation r, it creates an adjacency matrix M^r, where the value of $M^r_{i,j}$ is non-zero if triple (e_i, r, e_j) exists. Then, NeuralLP learns to reason by conducting matrix multiplication among different adjacency matrices.

Recently, ConvE [4] combines input entity s and relation r by 2D convolutional layers. It first reshapes the vectors of input s, r to 2D shapes, and then concatenates the two matrices for convolution operation. ConvE also describes a very simple model called InverseModel, which only learns inverse relations in a KG but achieves pretty good performance. Other models like [18,19] use extra resources that cannot be obtained from the original training data, such as text corpora or entity descriptions. We do not consider them in this paper.

To the best of our knowledge, all the existing models require both one entity s and its relation r provided to complete a triple. The proposed model in this paper is the first work that can predict the whole triples only given s.

3   Methodology

In this section, we first describe RNN and its multi-layer version. Then, we present DSKG, a variant of the multi-layer RNN specifically designed for KGs. To train DSKG effectively, we also propose a type-based sampling method. Finally, we introduce a method to enhance entity prediction with relation prediction.

3.1 RNN and Its Multi-layer Version

We start with the basic RNN model, which has only one RNN cell. Given a sequence $(x_1, \ldots, x_T)$ as input, the basic RNN model processes it as follows:

$$h_t = f(W_h h_{t-1} + W_x x_t + b),\tag{1}$$

where $f(\cdot)$ is an activation function, and W_h, W_x, b are parameters. h_t is the output hidden state at timestep t.

Multi-layer RNNs have shown promising performance on modeling complex hierarchical architectures in the NLP area [5]. By stacking multiple RNN cells, the complex features of each element can be hierarchically processed (see Fig. 1c). We write this below:

$$h_t^i = \begin{cases} f(W_h^i h_{t-1}^i + W_x^i x_t + b^i) & i = 0 \\ f(W_h^i h_{t-1}^i + W_x^i h_t^{i-1} + b^i) & i > 0 \end{cases},\tag{2}$$

where W_h^i, W_x^i, b^i are parameters for the i-th RNN cell. h_t^i is the hidden state of the i-th RNN cell at timestep t. Hence, each input element would be sequentially processed by each cell, which can be regarded as combining the concept of deep neural network (DNN) with RNN. In the end, we can use the hidden state of the last cell as output h_t at timestep t.

3.2 The Proposed Deep Sequential Model

Regarding triples in a KG as sequences enables us to model the KG with RNN. However, these length-3 sequences (i.e., triples) are quite special: entities and relations have very different characteristics and always crisscross in each triple. Therefore, we think that constructing respective multi-layer RNNs for entities and relations can help the model learn more complex structures. According to this intuition, we propose a KG-specific multi-layer RNN, which uses different RNN cells to process entities and relations respectively. As illustrated in Fig. 1d, using this architecture, the entire network is actually non-recurrent but still sequential. We write this structure as follows:

$$W_h^i = \begin{cases} E_h^i \\ R_h^i \end{cases} \quad W_x^i = \begin{cases} E_x^i \\ R_x^i \end{cases} \quad b^i = \begin{cases} b_E^i & x_t \in \mathbf{E} \\ b_R^i & x_t \in \mathbf{R} \end{cases}, \tag{3}$$

where $\mathbf{E}, \mathbf{R}$ denote the entity set and the relation set, respectively. We choose the current multi-layer RNN by the type of x_t, and then apply Eq. (2) for calculation.

3.3 Type-Based Sampling

Sampled softmax [6] is a very popular method for large label space classification. The underlying idea of this method is to sample a small number of negative classes to approximate the integral distribution. We write it as follows:

$$p_t = W_o h_t + b_o, \tag{4a}$$

$$L_t = -I(p_t, y_t) + \log \left(\sum_{\tilde{y} \in \{y_t\} \cup \mathbf{NEG}_t} \exp \left(I(p_t, \tilde{y}) \right) \right), \tag{4b}$$

where W_o, b_o are output weight matrix and bias. $I(p_t, y_t)$ returns the y_t-th value of p_t. We first use a fully-connected layer to convert output hidden state h_t to an unscaled probability distribution of label space, and then carry out the sampled softmax method to calculate the cross-entropy loss L_t. $\mathbf{NEG}_t$ denotes the negative set at timestep t. It is usually generated by a log-uniform sampler.

Furthermore, in KGs, label y_t also has its type. When y_t refers to an entity, it is meaningless to use negative relation labels for training, and vice versa. Therefore, we propose a customized sampling method that samples negative labels according to the type of y_t. We write it as follows:

$$\mathbf{NEG}_t = \begin{cases} Z(\mathbf{E}, n_e) & y_t \in \mathbf{E} \\ Z(\mathbf{R}, n_r) & y_t \in \mathbf{R} \end{cases}, \tag{5}$$

where $Z(\mathbf{E}, n_e)$ denotes the log-uniform sampler that samples the number of n_e labels from entity set $\mathbf{E}$. $Z(\mathbf{R}, n_r)$ is defined analogously. It is worth noting that, this sampler needs the labels in a lexicon sorted in descending order of frequency, thus we should also separately calculate the frequencies of entities and relations.

3.4 Enhancing Entity Prediction with Relation Prediction

Due to the input is length-3 triples, the model only minimizes two sub-losses for each triple. Given a triple (s, r, o), the model learns to predict r based on s, and to predict o based on $s \rightarrow r$. We propose a method that can leverage relation prediction for enhancing entity prediction. In Sect. 5.1, the experimental analysis proves that learning to predict relations is helpful for entity prediction.

Reversing relations is a commonly-used method to enable KG completion models to predict head and tail entities in an integrated fashion [10,14]. Specifically, for each triple (s, r, o) in the training set, a reverse triple (o, r^-, s) is constructed and added into the training set. Thus, a model can predict tail entities with input $(s, r, ?)$, and predict head entities with $(o, r^-, ?)$.

Previous models for KG completion need s, r to predict o. However, the ability of predicting relations enables our model to evaluate the probability distribution of reverse relations for each entity. For example, given an entity e_j, if the probability of $e_j \rightarrow r^-$ is very close to zero, then we can speculate that e_j does probably not have the relation r^-. In other words, e_j is not an appropriate prediction for $(s, r, ?)$. We formulate this by the following equation:

$$p'_{(s,r,?)} = (p_{(\mathbf{E},r^-)})^{\alpha} p_{(s,r,?)}, \tag{6}$$

where $p_{(\mathbf{E},r^-)}$ denotes the probability vector of r^- for all entities, and we calculate its element-wise power of α. We set $\alpha < 1$, since we want to alleviate the influence of such inaccurate prediction results. $p_{(s,r,?)}$ denotes the original probability vector of $(s, r, ?)$. For example, assume that the original entity probability vector is $(0.25, 0.25, 0.25)$. If we set $\alpha = \frac{1}{3}$, a reverse relation probability vector $(0.001, 0.8, 0.9)$ would be refined to $(0.1, 0.93, 0.97)$. By element-wise multiplication of the original entity probability vector and the refined reverse relation probability vector, we have $(0.025, 0.233, 0.243)$, which slightly affects those entities with high probabilities of r^-, but seriously penalizes those entities with near-zero probabilities. Consequently, the differences between entity probabilities are enlarged to help predict entities more accurately.

4 Experiments

4.1 Datasets and Experiment Settings

We implemented our model with TensorFlow and conducted a series of experiments on three datasets: FB15K [3], WN18 [3] and FB15K-237 [14]. Recent studies observed that FB15K and WN18 contain many inverse triple pairs, e.g., (USA, contains, NewYorkCity) in the test set and (NewYorkCity, containedby,

Table 1. Entity prediction results on two benchmark datasets

Models	FB15K				WN18			
	Hits@1	Hits@10	MRR	MR	Hits@1	Hits@10	MRR	MR
TransE[†] [3]	30.5	73.7	45.8	71	27.4	94.4	57.8	431
TransR[†] [11]	37.7	76.7	51.9	84	54.8	94.7	72.6	415
PTransE[†] [10]	63.8	87.2	73.1	59	87.3	94.2	90.5	516
DISTMULT [20]	54.6	82.4	65.4	97	72.8	93.6	82.2	902
NLFeat [14]	-	87.0	**82.1**	-	-	94.3	94.0	-
ComplEx [15]	59.9	84.0	69.2	-	93.6	94.7	94.1	-
NeuralLP [21]	-	83.7	76.0	-	-	94.5	94.0	-
ConvE [4]	67.0	87.3	74.5	64	93.5	95.5	94.2	504
InverseModel [4]	74.3	78.6	75.9	1,563	75.7	**96.9**	85.7	602
DSKG (cascade)	64.9	87.7	73.0	151	93.9	95.0	94.3	959
DSKG	**75.3**	**90.2**	80.9	**30**	**94.2**	95.2	**94.6**	**337**

"†" denotes the models executed by ourselves using the provided source code, due to some metrics were not used in literature.

"-" denotes the unknown results, due to they were unreported in literature and we cannot obtain/run the source code.

USA) in the training set [14]. By detecting the subjects and objects of *contains, containedby*, this inverse pair can be easily confirmed. So, the answer of (*USA, contains, ?*) is exposed. Even a very simple model that concentrates on these inverse relations can achieve state-of-the-art performance for many metrics [4]. Note that inverse triples are totally different from reverse triples. In our experiments, we more focus on FB15K-237, which was created by removing the inverse triples in FB15K. The detailed statistical data are listed in Table 3. We used the Adam optimizer [9] and terminated training when the results on the validation data were optimized. For each dataset, we used the same parameters as follows: learning rate $\lambda = 0.001$, embedding size $k = 512$ (initialized with the xavier initializer), and batch size $n_B = 2,048$. We chose the LSTM cells to implement the multi-layer RNNs and added the output dropout layer with keep probability $p_D = 0.5$ for each cell. The main results reported in this section is based on the 2-layer DSKG model. We will show parameter analysis in Sect. 5.

4.2 Entity Prediction

Following [3,4,14] and many others, four evaluation metrics were used: (i) the percentage of correct entities in ranked top-1 (Hits@1); (ii) in ranked top-10 (Hits@10); (iii) mean reciprocal rank (MRR); and (iv) mean rank (MR). Furthermore, we adopted the filtered rankings [3], which means that we only keep the current testing entity during ranking. Due to DSKG is capable of predicting relations only given one entity, we reported the so-called "**cascade**" results. Given a testing triple (s, r, o), DSKG first predicts the relations of $(s, ?)$ to obtain

Table 2. Entity prediction on FB15K-237

Models	Hits@1	Hits@10	MRR	MR
TransE[†]	13.3	40.9	22.3	315
TransR[†]	10.9	38.2	19.9	417
PTransE[†]	21.0	50.1	31.4	299
DISTMULT	15.5	41.9	24.1	254
NLFeat	-	41.4	27.2	-
ComplEx	15.2	41.9	24.0	248
NeuralLP	-	36.2	24.0	-
ConvE	23.9	49.1	31.6	246
InverseModel	0.4	1.2	0.7	7,124
DSKG (cascade)	20.5	50.1	30.3	842
DSKG	**24.9**	**52.1**	**33.9**	**175**

Table 3. Dataset statistics

Datasets	FB15K	WN18	FB15K-237
#Entities	14,951	40,943	14,541
#Relations	1,345	18	237
#Train	483,142	141,442	272,115
#Valid.	50,000	5,000	17,535
#Test	59,071	5,000	20,466

the rank of r, and then predicts the entities of $(s, r, ?)$ to obtain the rank of o. Finally, these two ranks are multiplied for comparison (i.e., the worst rank).

The experimental results on FB15K and WN18 are illustrated in Table 1. Because these two datasets contain many inverse triples, InverseModel, which only learns inverse relations, still achieved competitive performance. Additionally, we can observe that DSKG outperformed the other models for many metrics. Particularly, DSKG achieved the best performance for Hits@1, which showed that DSKG is quite good at precisely learning to predict entities. Even we evaluated DSKG in the cascade way, it still achieved comparable results.

Table 2 shows the entity prediction results on FB15K-237. We observed that: (1) The performance of all the models slumped. Specifically, InverseModel completely failed on this dataset, which reveals that all the models cannot directly improve their performance by using inverse relations any more. (2) DSKG significantly outperformed the other models for all the metrics. DSKG (cascade) also achieved state-of-the-art performance for some metrics (e.g., Hits@10).

4.3   Triple Prediction

DSKG is capable of not only predicting entities, but also predicting the whole triples. To evaluating the performance of DSKG on predicting triples directly, we constructed a beam searcher with a large window size. There also exist some complex methods that can improve sequence prediction performance [8]. Specifically, the model was first asked to take all the entities as input to predict relations, and then the top-100K $(entity, relation)$ pairs were selected to construct the incomplete triples like $(s, r, ?)$. Next, the model took these incomplete triples as input to predict their tail entities. Finally, we chose the top-1M triples as output, and sorted them in descending order for evaluation.

We used precision to assess these output triples. Let $\mathbf{S}_{out}^n$ denote the set of top-n output triples, $\mathbf{S}_{corr}$ denote the set of all correct triples (including the

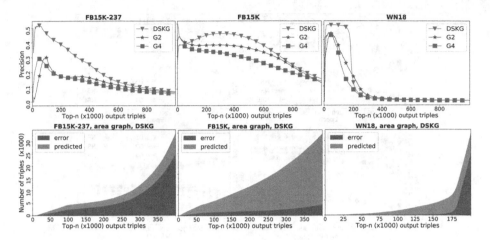

Fig. 2. Triple prediction results on three datasets

testing, validation and training sets) for a KG, and $\mathbf{S}_{pred}$ denote the set of predicted triples (including the testing and validation sets). The precision p_n w.r.t. top-n output triples is calculated as follows:

$$n_{corr} = |\mathbf{S}_{out}^n \cap \mathbf{S}_{corr}|, \quad n_{pred} = |\mathbf{S}_{out}^n \cap \mathbf{S}_{pred}|,$$

$$n_{error} = |\mathbf{S}_{out}^n| - n_{corr}, \quad p_n = \frac{n_{pred}}{n_{pred} + n_{error}}, \tag{7}$$

where $n_{corr}, n_{pred}, n_{error}$ denote the correct, predicted, and error numbers in $\mathbf{S}_{out}^n$, respectively. As a result, we can draw the curve of p_n in terms of n.

We conducted experiments on the three datasets, and compared DSKG with two general models: **G2** and **G4**. G2 is a general 2-layer RNN model (see Fig. 1c). G4 is a general 4-layer RNN model, as DSKG uses four different RNN cells. All the features (sampler, dropout, etc.) in DSKG were also applied to them.

As shown in the left column of Fig. 2, DSKG significantly outperformed G2 and G4 on all the datasets, especially for FB15K-237. Also, G4 performed worse than G2. This may be due to that deeper networks and more parameters make the entity and relation embeddings improperly trained. The right column of Fig. 2 shows the detailed triple prediction proportions of DSKG. DSKG predicted more than 2,000 correct triples with precision 0.47 (top-100K) on FB15K-237. On the other two easier datasets, DSKG performed better. It correctly predicted 34,155 triples on FB15K with precision 0.87 (top-400K) and 5,037 on WN18 with precision 0.91 (top-170K). Note that the precision of DSKG rapidly sharply dropped on WN18 in the end, because WN18 only has 10,000 triples to predict, while DSKG already output all the triples that it can predict.

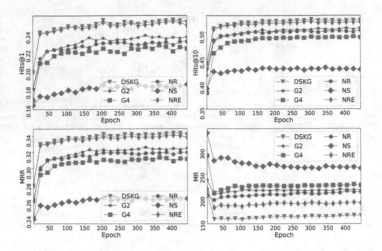

Fig. 3. Results of alternative models on FB15K-237

5   Analysis

5.1   Comparison with Alternative Models

To analyze the contribution of each part in DSKG, we developed a series of sub-models only containing partial features:

- **NR**. DSKG without using the relation loss in training. We constructed this model to assess the effect of minimizing relation loss to entity prediction.
- **NS** (non-sequential). We used four fully-connected layers (ReLU as the activation function) to replace RNN cells in DSKG, and cut down the connections between relation layer and entity layer. In the end, we added a dense layer to combine the output of these two layers. This model also shares the other features of DSKG (dropout, sampler, etc.). We constructed it to investigate the effect of sequential characteristic.
- **NRE**. DSKG without using the enhancement method (Sect. 3.4).

Figure 3 shows the performance of these models as well as G2 and G4 on the validation set of FB15K-237. From the results, we observed that:

- Sequential characteristic is a key point to DSKG. Comparing DSKG with NR and NS, we can find that: (1) although NR kept the sequential structure, it still performed worse than DSKG, since NR did not learn to predict relations; (2) NS did not use the sequential structure and not learn to predict relations. Hence, it obtained the worst result in Fig. 3.
- The KG-specific multi-layer RNN architecture significantly improved the performance. DSKG outperformed G2 and G4 for all the metrics on FB15K-237, even it did not use the enhancement from relation prediction. Note that, in Sect. 4.3, we have already shown that DSKG performed better than G2

Table 4. Entity prediction results with varied layer numbers on FB15K-237

No.	DSKG		DSKG (cascade)	
	Hits@1	Hits@10	Hits@1	Hits@10
1	24.6	51.2	20.3	48.8
2‡	**24.9**	**52.1**	**20.5**	**50.1**
3	24.6	51.4	19.8	49.3
4	24.1	50.3	18.6	48.1

"‡" denotes the main results reported in Sect. 4.

Table 5. Entity prediction results with varied embedding sizes on FB15K-237

Size	DSKG		DSKG (cascade)	
	Hits@1	Hits@10	Hits@1	Hits@10
512‡	**24.9**	**52.1**	**20.5**	**50.1**
256	24.8	**52.1**	19.9	49.8
128	24.5	51.5	19.0	48.5
64	23.1	48.6	17.1	45.1

and G4 on triple prediction. Therefore, the architecture used in DSKG has a better capability to model KGs than the general multi-layer RNN models.

– The relation enhancement method further refined the entity prediction results. DSKG always performed better than NRE, especially for MR, since it can directly eliminate many incorrect entities.

5.2 Influence of Layer Number

We conducted an experiment to analyze the influence of layer number to entity prediction. As shown in Table 4, when we increased the layer number from 1 to 4, DSKG cannot continuously improve the performance. The 4-layer model also took more time than the 2-layer model for each epoch (18.7 s vs. 13.4 s). Thus, we chose the 2-layer DSKG as the main reported version, which achieved the best and was trained quite fast (about one hour using a GTX 1080Ti).

5.3 Influence of Embedding Size

DSKG is a parameter-efficient model. Table 5 shows the entity prediction results of DSKG with varied embedding sizes. When we decreased the embedding size to 128, DSKG can still achieve state-of-the-art performance on FB15K-237. When the embedding size was set to 64, which is a very small value, it was also competitive. For DSKG (cascade), due to involving the relation prediction results, the performance decreased more severely, but still acceptable. We chose the embedding size 512 as the main reported version to obtain the best performance.

6 Conclusion and Future Work

In this paper, we proposed a new model to use a KG-specific multi-layer RNN to model triples in a KG as sequences. Our experimental results on three different datasets showed that our models achieved promising performance not only on the traditional entity prediction task, but also on the new triple prediction task. For the future work, we plan to explore the following two directions:

- Integrating the attention mechanism [1] in our model. While the attention mechanism has shown its power in the NLP area, applying this mechanism to KG completion has not been well studied. In the future, we want to extend DSKG with this mechanism to improve its inference ability.
- Using a provided KG to complete another KG. Recently, several methods start to leverage extra textual data for improving KG completion. However, textual data are written in natural language. Due to the ambiguity and heterogeneity, they may bring mistakes in prediction. Therefore, we think that taking existing KGs as another kind of extra data may improve performance.

Acknowledgements. This work was supported by the National Natural Science Foundation of China (Nos. 61872172 and 61772264).

References

1. Bahdanau, D., Cho, K., Bengio, Y.: Neural machine translation by jointly learning to align and translate. In: ICLR, San Diego, USA (2015)
2. Bollacker, K.D., Evans, C., Paritosh, P., Sturge, T., Taylor, J.: Freebase: a collaboratively created graph database for structuring human knowledge. In: SIGMOD, pp. 1247–1250. ACM, Vancouver (2008)
3. Bordes, A., Usunier, N., Garcia-Durán, A., Weston, J., Yakhnenko, O.: Translating embeddings for modeling multi-relational data. In: NIPS, pp. 2787–2795. Curran Associates, Inc., Lake Tahoe (2013)
4. Dettmers, T., Minervini, P., Stenetorp, P., Riedel, S.: Convolutional 2D knowledge graph embeddings. In: AAAI. AAAI, New Orleans (2018)
5. Hermans, M., Schrauwen, B.: Training and analysing deep recurrent neural networks. In: NIPS, pp. 190–198. Curran Associates, Inc., Lake Tahoe (2013)
6. Jean, S., Cho, K., Memisevic, R., Bengio, Y.: On using very large target vocabulary for neural machine translation. In: ACL, pp. 1–10. ACL, Beijing (2015)
7. Józefowicz, R., Vinyals, O., Schuster, M., Shazeer, N., Wu, Y.: Exploring the limits of language modeling. In: ICLR, San Juan, Puerto Rico (2016)
8. Keneshloo, Y., Shi, T., Ramakrishnan, N., Reddy, C.K.: Deep reinforcement learning for sequence to sequence models. CoRR abs/1805.09461 (2018)
9. Kingma, D.P., Ba, J.: Adam: a method for stochastic optimization. In: ICLR, San Diego, USA (2015)
10. Lin, Y., Liu, Z., Luan, H.B., Sun, M., Rao, S., Liu, S.: Modeling relation paths for representation learning of knowledge bases. In: ACL, pp. 705–714. ACL, Beijing (2015)
11. Lin, Y., Liu, Z., Sun, M., Liu, Y., Zhu, X.: Learning entity and relation embeddings for knowledge graph completion. In: AAAI, pp. 2181–2187. AAAI, Austin (2015)
12. Miller, G.A.: WordNet: an electronic lexical database. Commun. ACM **38**(11), 39–41 (1995)
13. Nickel, M., Tresp, V., Kriegel, H.P.: A three-way model for collective learning on multi-relational data. In: ICML, pp. 809–816. Omnipress, Bellevue (2011)
14. Toutanova, K., Chen, D.: Observed versus latent features for knowledge base and text inference. In: CVSC, pp. 57–66. ACL, Beijing (2015)
15. Trouillon, T., Welbl, J., Riedel, S., Gaussier, É., Bouchard, G.: Complex embeddings for simple link prediction. In: ICML, New York City, USA, pp. 2071–2080 (2016)

16. Vinyals, O., Kaiser, L., Koo, T., Petrov, S., Sutskever, I., Hinton, G.E.: Grammar as a foreign language. In: NIPS, pp. 2773–2781. Curran Associates, Inc., Montréal (2015)
17. Wang, Z., Zhang, J., Feng, J., Chen, Z.: Knowledge graph embedding by translating on hyperplanes. In: AAAI, pp. 1112–1119. AAAI, Québec City (2014)
18. Xiao, H., Huang, M., Meng, L., Zhu, X.: SSP: Semantic space projection for knowledge graph embedding with text descriptions. In: AAAI, pp. 3104–3110. AAAI, San Francisco (2017)
19. Xie, R., Liu, Z., Jia, J., Luan, H., Sun, M.: Representation learning of knowledge graphs with entity descriptions. In: AAAI, pp. 2659–2665. AAAI, Phoenix (2016)
20. Yang, B., Yih, W., He, X., Gao, J., Deng, L.: Embedding entities and relations for learning and inference in knowledge bases. In: ICLR, San Diego (2015)
21. Yang, F., Yang, Z., Cohen, W.W.: Differentiable learning of logical rules for knowledge base reasoning. In: NIPS, pp. 2316–2325. Curran Associates, Inc., Long Beach (2017)

Pattern Learning for Chinese Open Information Extraction

Yang Li[1]([✉]), Qingliang Miao[1], Tong Guo[1], Ji Geng[2], Changjian Hu[1], and Feiyu Xu[1]

[1] Lenovo, Building H, No.6, West Shangdi Road, Haidian District, Beijing, China
{liyang54,miaoql1,guotong1,hucj1,fxu}@lenovo.com
[2] UESTC, North JianShe Road, Chenghua District, Chengdu, China
jgeng@uestc.edu.cn

Abstract. Open Information Extraction systems, such as ReVerb, OLLIE, Clause IE, OpenIE 4.2, Sanford OIE, and PredPatt, have attracted much attention on English OIE. However, few studies have been reported on OIE for languages beyond English. This paper presents a Chinese OIE system PLCOIE to extract binary relation triples and N-ary relation tuples from Chinese documents. Our goal is to learn general patterns that is composed of both dependency parsing roles and parts of speech from large corpus, and the learned patterns are used to extract relation tuples from documents. In addition, this paper alleviates trans-classed word issue and light verb construction issue. PLCOIE can extract binary relation triples as well as N-ary relation tuples, and experiments on four real-world data sets show that the results are more precise than state-of-the-art Chinese OIE systems, which indicate that PLCOIE is feasible and effective.

Keywords: Information extraction · Trans-classed word · LVC
Logistic regression

1 Introduction

Open Information Extraction (OIE) is the task of extracting relation tuples (binary relation triples and N-ary relation tuples) from massive open-domain corpora. Comparing with traditional information extraction methods, OIE techniques can greatly reduce the dependence on expert-defined rules and relation types when building a knowledge base system, thus making it possible for people to deal with Web-scale corpora, and automatically tack an unbounded number of relations. The Turing center of Washington publishes the first OIE system TextRunner [1].

OIE system has drawn significant research interests recently. However, only a few studies have been made on Chinese, and the research progress is significantly behind the average pace. This is mainly due to the fact that Chinese language is uniquely different from English, most scientifically proven IE methods can't

© Springer Nature Singapore Pte Ltd. 2019
J. Zhao et al. (Eds.): CCKS 2018, CCIS 957, pp. 78–90, 2019.
https://doi.org/10.1007/978-981-13-3146-6_7

simply be transplanted "as-is" from English to Chinese. For instance, the Hearst patterns has been proven very effective in pattern-based information extraction tasks in English, however it is difficult to find a similar solution in Chinese. Some previous work has been done to build some useful patterns for relation extraction on Chinese corpus, but the achievement is still limited compared with the state-of-the-art OIE systems in English domain [2].

The arguments and research findings to be discussed in this paper focus on the problems involved in developing an open-domain relation extraction model for Chinese information extraction tasks. More specifically, we consider two issues in this paper.

The first is the trans-classed words issue, also known as the multi-tagging words problem, which means that many Chinese words can be used for different grammatical functions, that could cause significant effects on the precision of the POS (Part of Speech) tagging results [3,4]. For instance, Ex1 of Table 1 shows that the trans-classed word " 选举(election/elect)" has different POS tags in sentence 1 and sentence 2 without form inflection. In Chinese, the trans-classed words issue is a common linguistical phenomenon. In order to recognize and extract all of the relation tuples correctly from a sentence, one needs to carefully resolve the trans-classed words issue carefully, and make sure these words are accurately tagged according to their context.

The second is the light verb construction (LVC) issue. LVC is a multi-word expression composed of a light verb and a noun, with the noun carrying the semantic content of the predicate [5]. The light verb contains less information of its own, but can possess more information combining with a noun [6]. For example, "made a deal with"is a LVC, where"make" is a light verb and an uninformative relation phrase. Improper handling of LVC can cause uninformative extractions. Reverb uses the syntactic constraint, namely, relation phrase should include nouns, to solve the issue [5].

The contribution of this paper is three-folded. First, we introduce the idea of using the syntactic dependency parsing techniques to solve the trans-classed words issue, so as to further improve the performance of our pattern learning algorithm. Second, we propose a novel method for solving the LVC issue by extract N-ary tuple based on the result of dealing with trans-classed words issue. Third, we propose a new model called PLCOIE(Pattern Learning for Chinese Open Information Extraction), which learns 2430 high quality general patterns, for automatically extracting relation tuples from open domain Chinese corpora. The PLCOIE model can be used to extract both binary relation triples and N-ary relation tuples. The performance of the PLCOIE is carefully evaluated, comparing with state-of-the-art Chinese OIE systems, on four real world data sets. We publicize learned patterns at https://github.com/lbda1/PLCOIE.

2   Related Work

In recent years, a batch of excellent OIE systems are proposed. Many excellent OIE systems have been put forward. These works can be grouped according to the level of sophistication of the NLP techniques they rely upon: shallow parsing, deep NLP [7].

The representative systems using shallow parser are TextRunner and ReVerb. TextRunner uses a CRF model encoding POS and NP-chunked features to extract relation tuples [1]. ReVerb extracts a verb phrase that meets syntactic and lexical constraints [5]. A classifier is trained by shallow syntactic features to give a confidence score to each extracted triples. However, most incorrect tuples are caused by mismatch between relation phrase and arguments (ARGs).

Deep NLP based OIE systems rely on semantic role labeling (SRL) or dependency parsing to extract relation tuples. Systems using dependency parsing or SRL include DOIE, OLLIE, ClausIE, OpenIE 4.2, Stanford OIE and PredPatt. DOIE uses dependency path to identify the clauses of each sentence. Then predefined extraction rules are applied on the clause to extract binary relation triples [8]. OLLIE first extracts high confidence seed tuples and then learns syntactic, semantic and lexical patterns [9]. ClausIE first finds clauses in a given sentence. Subsequently it identifies the type of each clause according to the grammatical function of its constituents. Finally, binary relation triples are extracted using clause dependent patterns [10]. Stanford OIE, which is based on dependency parsing, yields a number of independent clauses of a sentence. Then each clause produces a set of entailed shorter utterances, and segments the ones that match an atomic pattern into a relation triple [11]. PredPatt provides a syntactic dependency annotation standard that can be used consistently across many languages including Spanish, Portuguese, Galician and English [12]. Some OIE systems also use SRL, however this kind of system is computationally expensive [13]. OpenIE 4.2[1] is the successor to OLLIE which uses similar argument and relation expansion heuristics to create OIE extractions from SRL frames.

Current OIE systems mainly focus on English while few studies have been reported on languages beyond English especially on Chinese. The representative Chinese OIE systems are CORE, ZORE and Logician. All of them are based on dependency parsing. At first, CORE adopts CKIP parser to analyze syntactic structure. Then it just recognizes "Head" labeled verb as relation. Finally, the noun phrases preceding/preceded by the relational chunk are regarded as the candidates of the head's arguments [2]. It can be regarded as using POS tagging and dependency parsing rules instead of semantic patterns. ZORE automatically identifies relation candidates from parsed dependency trees, and extracts relation with their semantic patterns iteratively through a novel double propagation algorithm [14]. ZORE's pattern focuses on specific relation phrase, instead of general relation phrase [14]. The following is an instance of ZORE's pattern: **nsubj(Af)_Pred(毕业)_prep(于)_pobj-NN(Di)**, which is fixed on the relation phrase 毕业(graduate). ZORE solved LVC issue by pre-constructed lexicon and verbobject structures that was obtained through statistics in a large-scale corpus. Our system PLCOIE copes with LVC issue using an easier way extracting N-ary tuples based on binary relation triples. Logician adopts Sequence-to-Sequence neural network model, which uses dependency tree to calculate gated depen-

---

[1] https://github.com/allenai/openie-standalone.

dency attention, to extract relation tuples [15]. These three systems do not deal with the trans-classed word issue.

3   Issues and Solutions

In this section, we introduce two linguistic phenomenons called trans-classed word and LVC. Both of them have a great influence on OIE system.

3.1   Trans-Classed Word

The trans-classed words consist of the same characters and are near-synonymous, but they have different grammatical functions [3, 4].

Table 1. Examples of trans-classed word.

Ex1. Verb-to-Noun Inflection
Trans-classed word: 选举(election/elect)
Sentence 1:公民(citizen)通过(via)投票(vote)选举(elect)出(out)总统(president)
POS in Sentence 1: verb
Sentence 2:奥巴马(Obama)赢得(win)总统(president)选举(election)
POS in Sentence 2: noun
Ex2. Adverb-to-Noun Inflection
Trans-classed word:秘密(secretly/secret)
Sentence1:他(He)秘密的(secretly)潜伏(lurk)在(at)敌营(enemy camp)
POS in Sentence 1: adverb
Sentence2:我(I)知道(know)你的(your)秘密(secret)
POS in Sentence 2: noun

In Chinese, some words have no form inflection in different parts of speech so trans-classed word is quite common. In this paper, we are mainly concerned with Verb-to-Noun Inflection and Adjective-to-Noun Inflection. And Table 1 shows corresponding examples. It is particularly worth identifying nouns accurately in sentences, because nouns are the main components of arguments of relation tuples. Such accurate identification facilitates all procedures in OIE that involve nouns such as noun phrase(NP) arguments extraction and LVC processing.

To address this question, we analyze the trans-classed words in 300 Tencent News sentences, among which, 175 are found with wrong POS tags. The statistical result indicates that this phenomenon is common in Chinese. Through careful analysis we have two observations and corresponding solutions as described later. For each trans-classed word in 300 Tencent News sentences, we check whether it satisfies observations. At last, 78 sentences satisfy the Observation 1, and 135 satisfy the Observation 2. The data indicates that these observations have broad coverage.

Observation 1: If the POS tag of a word is v(verb) and its dependency parsing role is ATT(attribute) then this word always acts as the semantics of noun. For instance, "驻(station)" in Fig. 1.

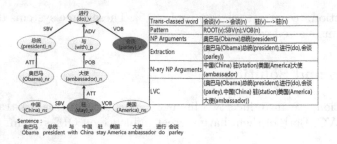

Trans-classed word	会谈(v)--->会谈(n) 驻(v)--->驻(n)
Pattern	ROOT(v);SBV(n);VOB(n)
NP Arguments	奥巴马(Obama)总统(president)
Extraction	(奥巴马(Obama)总统(president),进行(do),会谈(parley))
N-ary NP Arguments	中国(China) 驻(station)美国(America)大使(ambassador)
LVC	(奥巴马(Obama)总统(president),进行(do),会谈(parley),中国(China) 驻(station)美国(America)大使(ambassador))

Fig. 1. The dependency tree of the sentence.

Observation 2: The POS of a word is not a noun and its dependency parsing role is VOB/POB/IOB which means the word acts as object in the sentence. Besides, there is no another word with dependency parsing role VOB/POB/IOB that directly or via a preposition links to the word. Then the word acts as the semantics of noun. For instance, "会谈(parley)" in Fig. 1.

Based on these two observations, we use Zpar dependency parsing [16] result to identify and correct wrong trans-classed word's POS tag. In particular, if there is a word in the whole dependency tree which fits one of observations then we will replace its POS tag with n(noun). After this step, a more precise POS tags are prepared for the next step.

3.2 Light Verb Construction

LVC is a multi-word expression composed of a light verb and a noun, with the noun carrying the semantic content of the predicate [5]. In Chinese, LVC is frequent and should be handled properly in order to ensure that the relation tuples are informative. However, the syntactic constraints in Reverb can't be transferred to Chinese directly, because the word orders of English and Chinese are quite different [14]. ZORE solved the LVC issue by searching in a pre-constructed lexicon for verb-object structures that frequently co-occur with related preposition in a large-scale corpus. This method enables triples to become informative. For instance, ZORE can extract the relation tuple(奥巴马(Obama)总统(president), 进行(do)会谈(talk), 大使(ambassador)) from the sentence in Fig. 1 which treats "进行(do)会谈(talk)" as a whole relation phrase.

By observing N-ary relation tuple (奥巴马(Obama)总统(president), 进行(do), 会谈(talk), 大使(ambassador)) and binary relation triple (奥巴马(Obama)总统(president), 进行(do) 会谈(talk), 大使(ambassador)) these two tuples, we find that they have same information. The difference between these two relation tuples is the number of arguments. Through this example, we expand the extracted triples to N-ary tuples, and the LVC issue is solved directly. PLCOIE uses N-ary relation tuples to process LVC issue if a new noun phrase is connected to relation phrase by preposition. This method is much easier than ZORE's pre-constructed lexicon.

4 PLCOIE System

Our system tries to extract relation tuple candidates from sentences without predefined relation types. The sole input of PLCOIE is a Chinese corpus and the output is an extracted set of relation tuples with confidence scores. In this section, we describe the main components of PLCOIE in detail. The architecture is shown in Fig. 2. The system consists of four key modules: the first component is Trans-classed Word Process after which a more precise POS tags are prepared for the next step; the second component is Pattern Learning; the third component is Relation Candidates Extraction, during which we will extend same binary relation triple to N-ary relation tuples to solve LVC issue; the fourth component is Computing the Confidence. The confidence scores for every relation tuple is used to filter out the wrong tuples.

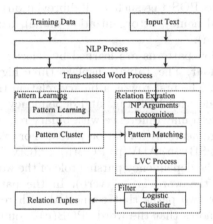

Fig. 2. Architecture of PLCOIE.

4.1 Pattern Learning

In this section, we introduce how to prepare substantial training data (binary triple and corresponding sentence pairs) and learn general patterns which are composed of POS tags and dependency parsing roles. There are two methods to extract training data. One method first extracts 10 thousand binary relation triples from Baidu encyclopedia[2] infobox and then searches, extracts corresponding sentences which contain both arguments of every triple from the web by distant supervised method. The other method is uses start-of-art Chinese OIE system to extract and filter substantial high confidence binary relation triples from 300 thousand Tencent News sentences. Finally we extract almost 50 thousand binary relation triple and corresponding sentence pairs.

---

[2] https://baike.baidu.com/.

In order to learn relation patterns, we need to extract the dependency path that connects two arguments and relation phrase for each binary relation triple from the corresponding sentence. The arguments of relation triples are noun phrases, hence we use the head word of arguments to represent it during learning relation patterns. When we learned a pattern from a sentence and at the same time the pattern's frequency is increased. Figure 1 shows an example of training data entry: binary relation triple (奥巴马(Obama) 总统(president), 进行(do), 会谈(parley)) and its corresponding sentence. First, we use 总统(president) the head word of the ARG1 to stand for it. Then we connect the ARG1's head word, Relation phrase, ARG2's head word in dependency parsing path. Finally we get SBV(n)-Root(v)-VOB(n) after dealing with trans-classed word followed by two observations of Sect. 3.1.

In order to make the patterns more effective, we group patterns according to POS tags to get more coarse-grained patterns. The detailed process is that we replace POS tag such as j(abbreviation), i(idiom), nr(person name), ns(geographical name), nt(organization name), nz(other proper noun), r(pronoun) with n(noun).

Finally, 2430 relation patterns are learned by selecting the patterns whose frequency more than four. The Table 2 displays three high-frequency patterns learned by our system. The first row is a binary relation triple for patterns(the second row) that encode with POS tags and dependency parsing roles. Table 2 indicates that the pattern may difference with binary relation triple due to pattern connects both arguments with relation phrase. For example in Table 2, the learned pattern "IC(v):SBV(n):ADV(p),POB(n)", "IC(v)" denotes the relation phrase, where "IC" is the dependency parsing role of the word and "(v)" denotes that the POS tag of the word is "v"(verb). In the pattern, "SBV(n)" and "ADV(p),POB(n)" are two paths that connected with relation word "IC(v)". "ADV(p),POB(n)" denotes that the word "POB(n)" connects to relation word via a intermediate word "ADV(p)". "(IC(v),SBV(n),POB(n))" is binary relation triple which can be extracted by the pattern. Because the pattern is the dependency path that connects two arguments and relation phrase so it may be longer than relation triples.

Table 2. Top three high-frequency pattern templates.

Relation triples	Patterns
(IC(v), SBV(n), VOB(n))	IC(v):SBV(n):VOB(n)
(ROOT(v), SBV(n), VOB(n))	ROOT(v):SBV(n):VOB(n)
(IC(v), SBV(n), POB(n))	IC(v):SBV(n):ADV(p),POB(n)

4.2   Relation Candidates Extraction

This section describes how these patterns are used to extract binary relation triples and solve the LVC issue by expanding to N-ary relation tuples in detail.

First, we match the patterns with the dependency parsing result of the sentence by algorithm closetString [17]. Second, head words of relation phrase and arguments are extracted by the matched pattern. Third we expand arguments to contain more information, since the head words contain less information. For expanding arguments, we merge the head words with noun of its dependency subtree. Finally, some extracted binary triples are expanded to N-ary tuples to avoid uninformative extractions if a new noun phrase is connected with relation phrase by preposition. The LVC issue is solved directly after expanding the extracted binary relation triples to N-ary tuples.

We also use the example of Fig. 1 to explain how to use the learned 2430 patterns to extract relation tuples. The figure shows that an example has trans-classed word issue("会谈(parley)" and "驻(station)") and LVC issue("进行(do)") at the same time. Firstly, PLCOIE is based on the result of NLP preprocessing and trans-classed words processing to select the matched pattern as showed in the second line of Fig. 1. Secondly, according to matched pattern, PLCOIE extracts the relation phrase("进行(do)"), head word of arguments("总统(president)" and "会谈(parley)") and then expand the argument to NP arguments " 奥巴马(Obama) 总统(president)". The extracted binary relation triple is showed in the fourth line of Fig. 1. Finally, because there is a new noun phrase(" 中国(China)驻(station)美国(America) 大使(ambassador)") which is connected with relation phrase("进行(do)") by preposition("与(with)"), we expand the binary triple to N-ary tuple as showed in the fifth line of Fig. 1.

4.3 Computing the Confidence

There are still some error candidate relation tuples of PLCOIE. In order to improve performance, PLCOIE trains a confidence function to filter error tuples with low confidence. A supervised logistic regression classifier is used produce a confidence score to each relation tuples, with features shown in the Table 3. "ArgumentCount"indicates the number of NP arguments. "Comma-Count" indicates the number of comma in sentence. "PatternFrequency" indicates the frequency number of the pattern which matches the tulple. "WordsCount" indicates the number of node in dependency tree of the sentence. "MaxDistanceS_NP_R"presents the maximal distance from all NP arguments to relation phrase in the sentence. "MaxDistanceDP_NP_R" denotes the maximal distance from all NP arguments to relation phrase in dependency tree. "MaxDistanceS_NPs" denotes the maximal distance between different NP arguments in sentence. The classifier learns a weight for every feature and all of these features are effective. The table shows the weights of all features trained on the Wiki-500 data set.

5 Experiments and Discussion

We conduct an experimental study to compare PLCOIE with relevant approaches on four data sets: Wiki-500, Sina-500, Tencent-500 and Simple-500.

Table 3. Features of the logistic regression classifier with weights on Wiki-500 data set.

Feature	Weight	Feature	Weight
MaxDistanceS_NP_R < 5	0.7584	5 ≤ MaxDistanceS_NP_R < 10	0.2204
10 ≤ MaxDistanceS_NP_R	−0.2414	5 ≤ MaxDistanceS_NPs < 10	−0.194
10 ≤ MaxDistanceS_NPs	0.2597	ArgumentCount = 2	0.7883
ArgumentCount =3	0.6477	ArgumentCount > 3	−1.042
CommaCount ≤ 1	−0.3785	CommaCount ≥ 2	−0.2274
PatternFrequency ≥ 1000	1.326	100 ≤ PatternFrequency < 1000	0.5521
10 ≤ PatternFrequency < 100	0.359	5 ≤ PatternFrequency < 10	0.8762
PatternFrequency < 5	−0.7193	MaxDistanceDP_NP_R > 2	−0.1569

Each data set contains 500 sentences, summarized in Table 4. Wiki-500, Sina-500 are used as experiment data in ZORE [14]. In order to evaluate whether PLCOIE is effective in long and complex sentence or not, we construct Tencent-500 that consists of long and complex Tencent News sentences. We also construct Simple-500 to evaluate whether PLCOIE is more effective on short data set. For Simple-500, the length of each sentence limited at 40.

Table 4. Four data sets used in this paper.

DataSet	Source	Sentences_Number	Tuples_ Number
Wiki-500	ZORE Wiki data	500	816
Sina-500	ZORE Sina News	500	1043
Tencent-500	Tencent News data	500	941
Simple-500	Web simple data	500	497

Three graduate students with background of NLP are recruited to judge and evaluate each extracted tuples independently. To construct the final gold standard, we adopt the following procedure. If the labeled result of three annotators are the same, we assign this agreed-upon label. If the labeled result is different, then three annotators discuss their assessment with each other in a face-to-face meeting. We then used their consensual assessment as the final label.

Based on the above gold standard, Precision, Recall, F-Measure is used to evaluate the effects of PLCOIE. They are defined as follows

$$\text{Precision} = \frac{c\_T}{e\_T} \qquad \text{Recall} = \frac{c\_T}{R} \qquad \text{F} - \text{Measure} = \frac{2 * Precision * Recall}{Precision + Recall}$$

where R is all relevant tuples of data set; e_T represents the set of tuples that are extracted by PLCOIE; c_T represents the correctly labeled tuples in e_T.

5.1   Comparing with ZORE

We compare PLCOIE with ZORE on four data sets. Since logistic confidence score is given by both PLCOIE and ZORE for every extracted relation tuple, we can measure the precision and recall of two systems with different confidence threshold.

Figure 3 shows the precision-recall curves of the PLCOIE and ZORE on four data sets by varying confidence threshold. Experimental results on four data sets show that PLCOIE performs better than ZORE. The PR Curves on Wiki-500, Sina-500, Tecen-500 all obviously decrease along as the increasing Recall, but the curve tends to flatten on Simple-500 data set. The reason is that relation tuples extracted by PLCOIE at simple sentences are more accurate. As showed in Fig. 3, precision almost reached 90% when recall is 70%.

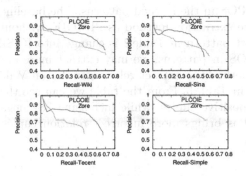

Fig. 3. Comparing performance of PLCOIE with ZORE on four data sets.

5.2   The Importance of Dealing with Trans-Classed Word

A more accurate POS tagging result is obtained via dealing with trans-class word issue followed by two observations of Sect. 3.1. And the statistics indicates that our solution has broad coverage. In this section, we compare PLCOIE with a baseline to highlight the effectiveness of solving trans-classed words. The baseline PLCOIE-NoTCW uses the same extraction method as PLCOIE, without dealing with trans-classed words.

Figure 4 shows the precision-recall curves of the systems PLCOIE and PLCOIE-NoTCW on Wiki-500 by varying the confidence threshold. The figure presents that PLCOIE has higher precision than PLCOIE-NoTCW at almost all different recall. Besides, PLCOIE has higher recall than PLCOIE-NoTCW at almost all different precision. In Fig. 4, when the recall is 0.48, the precision of PLCOIE is 0.82, 0.22 higher than PLCOIE-NoTCW. At the same time, when the precision is 0.65, the recall of PLCOIE is 0.69, 0.265 higher than PLCOIE-NoTCW.

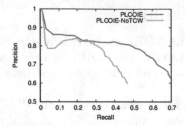

Fig. 4. Comparing performance of PLCOIE with PLCOIE-NoTCW on Wiki-500.

5.3 The Importance of Part of Speeach in the Pattern

In this section, we compare PLCOIE model with a baseline system to highlight the effectiveness of POS in the learned patterns. The baseline system PLCOIE-NoPos uses the same extraction method as PLCOIE, without POS in patterns. For example the first pattern in Table 2 is change into IC:SBV:VOB. Without the constraint of POS in patterns, we may extract more error relation tuples. We use these two methods to extract relation tuples on Wiki-500 data set and the result is shown in Table 5. From the table we can see that PLCOIE-NoPos only increases 103 positive tuples, while increasing 382 negative tuples. Obviously, PLCOIE-NoPos brings more incorrect relation tuples, greatly reducing the precision.

Table 5. Extracted tuples distribution of PLCOIE and PLCOIE-NoPos on Wiki-500.

Method	Relation tuples	Positive tuples	Negative tuples
PLCOIE	904	569	335
PLCOIE-NoPos	1389	672	717

Figure 5 shows the comparison of PLCOIE and PLCOIE-NoPos on Precision-Confidence and Fvalue-Confidence curves according to logistic confidence scores for every extracted relation tuples. In the precision-confidence curve of the figure, we can see that the precision of PLCOIE is far more better than baseline: 0.15 higher than that at the low confidence situation. With the confidence increasing, our result are still better than or equal to PLCOIE-NoPos. PLCOIE-NoPos yields many negative tuples, but with confidence increasing, the precision is better and better, even slightly better than PLCOIE sometimes. Error tuples caused by PLCOIE-NoPos are filtered by our logistic classifier, thus, it have high precision. In the Fvalue-Confidence curve of Fig. 5, we can see the F-value of two systems almost equal. It indicates that PLCOIE can make up the lost of recall by precision for the limitation of POS.

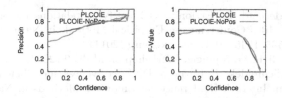

Fig. 5. Comparing performance of PLCOIE with PLCOIE-NoPos on Wiki-500.

5.4 Error Analysis

To better understand the limitations of PLCOIE and to improve our system, we perform a detailed analysis of its incorrect extractions and its missing extractions. We analysis and summarize the error types of incorrect relation tuples that are extracted by PLCOIE based on Wiki-500 data set. There are 66% incorrect relation tuples are caused by mismatching between arguments and relation phrase. Another main source of failure is incomplete argument phrase, which accounts for 29% of the total errors. The remaining errors may be caused by basic NLP processing such as segmentation, POS tagging as well as dependency parsing. We also summarize the reasons for missing relation tuples by PLCOIE. One common reason is segmentation and POS tagging errors, which accounts for 54% of the total errors. Another main source of failure is the limitation of patterns, which accounts for 34%.

6 Conclusion

This study proposes a system called PLCOIE to extract Chinese relation tuples. PLCOIE model automatically learns 2430 general and high quality relation patterns from large training corpus. Besides, our system deals with trans-classed word issue and LVC issue. For future research, we plan to scale the extracted relation tuples and add them into knowledge graph.

References

1. Banko, M., Cafarella, M.J., Soderland, S., Broadhead, M., Etzioni, O.: Open information extraction for the web. In: 20th International Proceedings on International Joint Conferences on Artificial Intelligence, pp. 2670–2676. University of Washington, Seattle (2007)
2. Tseng, Y.H., et al.: Chinese open relation extraction for knowledge acquisition. In: 14th International Proceedings on European Chapter of the Association for Computational Linguistics, pp. 12–16. ACL, Stroudsburg (2014)
3. Zhang, H., Zheng, J.: A study on consistency checking method of part-of-speech tagging for chinese corpora1. IJCLCLP **13**(2), 157–169 (2008)
4. Zhang, H., Zheng, J.H., Zhao, Y.: A classification-based algorithm for consistency check of part-of-speech tagging for Chinese corpora. J. Vis. Exp. Jove **pii(16)**, e722–e722 (2008)

5. Fader, A., Soderland, S., Etzioni, O.: Identifying relations for open information extraction. In: 16th Conference on Empirical Methods in Natural Language Processing, pp. 1535–1545. ACL, Stroudsburg (2011)
6. Butt, M.: The light verb jungle. Workshop on Multi 9 (2003)
7. Mesquita, F., Schmidek, J., Barbosa, D.: Effectiveness and efficiency of open relation extraction. In: 18th Conference on Empirical Methods in Natural Language Processing, pp. 225–252. ACL, Stroudsburg (2013)
8. Gamallo, P., Garcia, M., Ndez-Lanza, S.: Dependency-based open information extraction. In: 18th Joint Workshop on Unsupervised and Semi-Supervised Learning in NLP at the Conference of the European Chapter of the Association for Computational Linguistics, 10–18. ACL, Stroudsburg (2013)
9. Mausam, S.M., Bart, R., Soderland, S., Etzioni, O.: Open language learning for information extraction. In: 17th Joint Conference on Empirical Methods in Natural Language Processing and Computational Natural Language Learning. ACL, Stroudsburg (2012)
10. Del Gemulla, L., Corro, R.: Clausie: clause-based open information extraction, pp. 355–366 (2013)
11. Angeli, G., Premkumar, M.J.J., Manning, C.D.: Leveraging linguistic structure for open domain information extraction. In: 53th Association for Computational Linguistics and the 7th International Joint Conference on Natural Language Processing, pp. 344–354. ACL, Stroudsburg (2015)
12. White, A.S., et al.: Universal decompositional semantics on universal dependencies. In: 21th Proceedings of the 2016 Conference on Empirical Methods in Natural Language Processing, pp. 1713–1723. ACL, Stroudsburg (2016)
13. Christensen, J., Soderland, S., Etzioni, O., et al.: An analysis of open information extraction based on semantic role labeling. In: 6th Proceedings of the sixth international conference on Knowledge capture, pp. 113–120. ACM, New York (2011)
14. Qiu, L., Zhang, Y.: ZORE: a syntax-based system for Chinese open relation extraction. In: 19th Proceedings of the 2014 Conference on Empirical Methods in Natural Language Processing, pp. 1870–1880. ACL, Stroudsburg (2014)
15. Sun, M., Li, X., Wang, X., Fan, M., Feng, Y., Li, P.: Logician: a unified end-to-end neural approach for open-domain information extraction. In: 11th Eleventh ACM International Conference on Web Search and Data Mining. ACM, New York (2018)
16. Zhang, Y., Clark, S.: Syntactic processing using the generalized perceptron and beam search. Comput. Linguist. **37**(1), 105–151 (2011)
17. Li, M., Ma, B., Wang, L.: On the closest string and substring problems. J. ACM **49**(2), 157–171 (2002)

Adversarial Training for Relation Classification with Attention Based Gate Mechanism

Pengfei Cao[1,2], Yubo Chen[2], Kang Liu[1,2], and Jun Zhao[1,2(✉)]

[1] University of Chinese Academy of Sciences, Beijing, China
[2] Institute of Automation, Chinese Academy of Sciences, Beijing, China
caopengfei2018@ia.ac.cn, {yubo.chen,kliu,jzhao}@nlpr.ia.ac.cn

Abstract. In recent years, deep neural networks have achieved significant success in relation classification and many other natural language processing tasks. However, existing neural networks for relation classification heavily rely on the quality of labelled data and tend to be overconfident about the noise in input signals. They may be limited in robustness and generalization. In this paper, we apply adversarial training to the relation classification by adding perturbations to the input vectors in bidirectional long short-term memory neural networks rather than to the original input itself. Besides, we propose an attention based gate module, which can not only discern the important information when learning the sentence representations but also adaptively concatenate sentence level and lexical level features. Experiments on the SemEval-2010 Task 8 benchmark dataset show that our model significantly outperforms other state-of-the-art models.

Keywords: Relation classification · Adversarial training
Attention based gate mechanism

1 Introduction

The task of relation classification is to identify the semantic relations between the pair of nominal entities in a sentence, which can be defined as follows: given a sentence S with the annotated entity mentions e_1 and e_2, the goal is to predict the relation between e_1 and e_2 [10]. For instance, given the example input *"The $<e_1>$ burst $</e_1>$ has been caused by water hammer $<e_2>$ pressure $</e_2>$.",* the marked entities "burst" and "pressure" are of the relation Cause-Effect(e_1,e_2). Relation classification is an important step in natural language processing (NLP) applications that need to mine explicit facts from text such as information extraction [24], question answering [29] and knowledge base completion [3]. It has attracted much attention in recent years.

Traditional methods, based on either human-designed features [10] or kernels [2], have been proposed for relation classification. Although these methods have achieved high performance, however, these approaches inevitably suffer from two

© Springer Nature Singapore Pte Ltd. 2019
J. Zhao et al. (Eds.): CCKS 2018, CCIS 957, pp. 91–102, 2019.
https://doi.org/10.1007/978-981-13-3146-6_8

limitations. First, the extracted features or elaborately designed kernels are often derived from the output of existing NLP systems, which leads to the propagation of errors in the existing tools. Second, carefully handcrafted features perform poor on generalization due to the low coverage of different training datasets.

Recently, with the development of deep learning, neural networks have been introduced to relation classification. [21] introduced a recursive neural network model to learn compositional vector representations of sentences. [9] proposed a simple customization of recursive neural networks. [30] utilized convolutional neural networks (CNNs) to extract compositional semantics for relation classification. [28] applied long short term memory (LSTM) based recurrent neural networks (RNNs) along the shortest dependency path. [31] leveraged bidirectional LSTM (BLSTM) to capture complete and sequential information about all words. [32] introduced attention mechanism into relation classification to learn the "importance" distribution over the inputs.

Although great improvements have been achieved by neural network based methods, several problems still have not been well addressed. One problem is that there is only a very small amount of annotated data available and neural network based models are overconfident about the noise in input signals. Hence, these methods lack generalizability for unseen examples and robustness for adversarial examples which are generated by adding noise in the form of small perturbations to the original data [23]. The other problem is that they cannot efficiently exploit different level features. Most of neural network based methods only make use of sentence level features [32]. [30] utilized lexical level features by directly concatenating them with sentence level features. However, they cannot adaptively combine sentence level and lexical level features.

To address the above problems, we introduce adversarial training for relation classification, aiming to not only enhance the robustness for adversarial examples, but also improve generalization performance for original data. In addition, we propose attention based gate mechanism to capture important semantic information and dynamically combine sentence level and lexical level features. Specifically, we first create small perturbations based on gradient information and then add them to input embeddings. We employ BLSTM to extract sentence level features. After extracting features, we utilize attention mechanism to learn the important semantic information and adaptively combine sentence level and lexical level features with gate module. The contributions of this paper could be summarized as follows:

- We introduce an adversarial training strategy in bidirectional LSTM network for relation classification to improve the robustness and generalizability of the neural relation extraction model.
- A adaptive attention based gate mechanism is used to determine which part of sentence is more important and adaptively concatenate sentence level and lexical level features. To our best knowledge, it is the first work to dynamically combine sentence level and lexical level features with gate mechanism.

- We conduct experiments on the SemEval-2010 Task 8 benchmark dataset. The experimental results show that our model yields better performance than previous state-of-the-art models.

2 Related Work

Relation classification is an active area in NLP tasks, and there are many approaches have been introduced for relation classification. Apart from a few unsupervised clustering methods [4], the majority of work on relation classification has been supervised. In the supervised methods, relation classification is considered as a multi-class or multi-label problem. Generally, methods for relation classification mainly fall into three classes: feature based methods, kernel based methods and neural network based methods.

In the feature based methods, a set of high-level features such as part of speech tags, named entities and shortest dependency path are extracted from the output of an explicit linguistic preprocessing step [8] and fed into a classifier. Kernel based methods make use of various kernels, such as convolution tree kernels [18], subsequence kernels [16] and dependency tree kernels [2]. These traditional methods heavily depend on either carefully handcrafted features, often chosen on a trial-and-error basis, or elaborately designed kernels, which are often derived from pre-trained NLP tools. Therefore, such approaches inevitably suffer from the accumulations of errors and are limited in lexical generalization abilities for unseen words.

In recent years, deep neural networks have shown promising results. [21] proposed a recursive neural network model to capture the compositional vector representations of the sentence semantics. [30] introduced a deep convolutional neural network to extract sentence level features. In parallel, lexical features were extracted according to given entities. [20] tackled the relation classification task exploiting a convolutional neural network with a ranking loss function. [31] leveraged a bidirectional LSTM network to obtain sequential information of input sentence. Attention mechanism is well-known by its capability of learning the "importance" distribution over the inputs and demonstrates success in a wide range of tasks such as machine translations [1], question answering [11] and image captioning [27]. [32] proposed attention based neural networks to learn crucial information for relation classification.

Adversarial training [7] was originally introduced in computer vision [6] and has achieved great success. In the NLP community, [15] introduced adversarial training to text classification by adding perturbations on input vectors. Adversarial training utilizes the gradient information to determine the data perturbation. [26] proposed several data noise techniques for language model. [25] used adversarial training method to distant relation extraction task.

In this paper, we propose a simple yet effective architecture that enhances the robustness and generalization of model exploiting adversarial training and dynamically combines sentence level and lexical level features with adaptive attention based gate module.

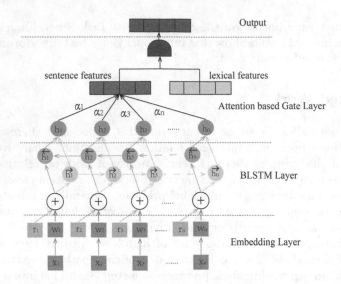

Fig. 1. The architecture of our proposed model. r_i denotes the adversarial perturbation of input embedding e_i.

3   Methodology

In this paper, we propose a novel neural network based model to improve robustness and generalization as well as dynamically combine sentence level and lexical level features. The architecture of our proposed model is shown in Fig. 1. The model mainly consists of three components: embedding layer, BLSTM layer and attention based gate layer. In the following sections, we will describe each part of our proposed model in detail.

3.1   Input Representation

Similar to other neural network models, the first step of our proposed model is to map discrete words into the distributed representations. To capture more syntactic and semantic information about every word, we exploit three kinds of information to construct input representations:

Word Embedding (WE). Given a sentence S consisting of n words $S = \{x_1, x_2, \ldots, x_n\}$, every word x_i is converted into a real-valued vector e_i. Given a word embedding matrix W_V of dimensionality $d_w \times |V|$, where V is a fixed-size input vocabulary and d_w is the dimensionality of word vector, we transform a word x_i into its word embedding e_i by mapping x_i to a column vector of W_V.

Part-of-Speech Embedding (PE). In our experiment, we use Stanford CoreNLP Toolkit to obtain the part-of-speech tags, and then initialize the embeddings randomly. The dimension of part-of-speech embedding is set 20.

Name Entity Embedding (NE). We also use Stanford CoreNLP Toolkit to obtain the name entities, and then initialize the embeddings randomly. The dimension of name entity embedding is set 10.

Finally, we concatenate the word embedding, part-of-speech embedding and named entity embedding of each word and denote it as input representation $w = [WE, PE, NE] \in \mathbb{R}^{d_e}$.

3.2   Bidirectional Long Short Term Memory

Long short term memory (LSTM) [12] is a variant of recurrent neural network (RNN) [5], which enables to address the gradient vanishing and exploding problems in RNN via introducing gate mechanism and memory cell. The LSTM with input gate i, forget gate f, output gate o and memory cell c works as follows at step t:

$$i_t = \sigma(W_i h_{t-1} + U_i w_t + b_i) \tag{1}$$
$$f_t = \sigma(W_f h_{t-1} + U_f w_t + b_f) \tag{2}$$
$$g_t = tanh(W_g h_{t-1} + U_g w_t + b_g) \tag{3}$$
$$c_t = f_t \odot c_{t-1} + i_t \odot g_t \tag{4}$$
$$o_t = \sigma(W_o h_{t-1} + U_o w_t + b_o) \tag{5}$$
$$h_t = o_t \odot tanh(c_t) \tag{6}$$

where $W_j \in \mathbb{R}^{d_h \times d_h}$, $U_j \in \mathbb{R}^{d_h \times d_e}$ and $b_j \in \mathbb{R}^{d_h}$ are trainable parameters, $j \in \{i, f, g, o\}$. σ denotes the sigmoid function and $\odot$ is element-wise product.

In order to incorporate information from both sides of sequence, we adopt BLSTM to extract features. At time step t, the hidden states of BLSTM can be expressed as follows:

$$\overrightarrow{h_t} = \overrightarrow{LSTM}(w_t, \overrightarrow{h_{t-1}}) \tag{7}$$
$$\overleftarrow{h_t} = \overleftarrow{LSTM}(w_t, \overleftarrow{h_{t+1}}) \tag{8}$$
$$h_t = \overrightarrow{h_t} \oplus \overleftarrow{h_t} \tag{9}$$

where $\overrightarrow{h_t}$ and $\overleftarrow{h_t}$ are the hidden states of the forward and backward LSTM, respectively. $\oplus$ is concatenation operation.

3.3   Attention Based Gate Mechanism

Attention mechanism has successfully been applied to a wide range of natural language processing tasks such as machine translations [1] and question answering [11]. In our work, we apply attention mechanism into relation classification task. The attention mechanism enables our model to determine which parts of the sentence are most influential for relation classification. The calculation procedure of attention mechanism is as follows:

$$m_i = W^T tanh(h_i) \tag{10}$$

$$\alpha_i = \frac{exp(m_i)}{\sum_{j=1}^{n} exp(m_j)} \tag{11}$$

$$h = tanh(\sum_{i=1}^{n} \alpha_i h_i) \tag{12}$$

where $h_i \in \mathbb{R}^{2d_h}$ denotes the hidden state of BLSTM at time step i, $W \in \mathbb{R}^{2d_h}$ is trainable parameter.

Lexical level features enable to provide important cues for deciding relations. In our work, we directly select the word embeddings of two marked entities and their left and right token as lexical level features denoted as $l \in \mathbb{R}^{6d_w}$. To exploit different level features, we adopt gate mechanism to dynamically combine sentence level and lexical level features. The gate mechanism works as follows:

$$G = \sigma(W_g(h \oplus l) + b_g) \tag{13}$$
$$f = G \odot h + (1 - G) \odot (M_l * l) \tag{14}$$

where $W_g \in \mathbb{R}^{2d_h \times (2d_h + 6d_w)}$ is weight matrix, $b_g \in \mathbb{R}^{2d_h}$ is bias term, and $M_l \in \mathbb{R}^{2d_h \times 6d_w}$ is conversion matrix.

3.4   Adversarial Training

In this setting, we use a fully connected softmax layer to predict label $\hat{y}$ from label set. The classifier is implemented using following equations:

$$p(y|S) = softmax(W_s f + b_s) \tag{15}$$
$$\hat{y} = \arg \max_{y} p(y|S) \tag{16}$$

Given T training examples $(x^{(i)}, y^{(i)})$, the loss function is the negative log-likelihood of the true class label:

$$L(\theta) = -\frac{1}{T} \sum_{i=1}^{T} log(y^{(i)}|x^{(i)}; \theta) \tag{17}$$

where θ denotes the parameters of model.

To improve the robustness and generalization of our proposed model, we introduce adversarial training for the classifier. Based on Eq. (17), we first generate small perturbations, and then add perturbations to input embeddings. Formally, the adversarial training can be written as follows:

$$L_{adv}(\theta) = -\frac{1}{T} \sum_{i=1}^{T} log(y^{(i)}|x^{(i)} + r_{adv}; \theta) \tag{18}$$

$$r_{adv} = -\epsilon g/\|g\|, \quad where \quad g = \nabla_x log p(y|x) \tag{19}$$

where r_{adv} is the perturbations of input embeddings. ϵ is a hyper-parameter. θ denotes model parameters.

In training phrase, we optimize the loss function defined as Eq. (18) using Adam algorithm [13].

4 Experiments

4.1 Dataset and Evaluation Metrics

We conducted our experiments on the SemEval-2010 Task 8 dataset [10], which contains 8000 sentences for training, and 2717 sentences for testing. Each sentence is annotated with each of nine types of annotated relations and an additional "Other" type. The nine types are: Cause-Effect, Component-Whole, Content-Container, Entity-Destination, Entity-Origin, Instrument-Agency, Member-Collection, Message-Topic and Product-Producer, while the relation type "Other" indicates that the relation expressed in the sentence is not in the above nine types. The former nine relation types are directed, whereas the "Other" type is undirected. For example, Cause-Effect(e_1,e_2) and Cause-Effect(e_2,e_1) are considered two distinct relations, so the total number of relation types is 19. We evaluate our model by using the SemEval-2010 Task 8 official scorer, which computes the Macro-F1 score for the nine relation types (excluding "Other") and takes the directionality into consideration.

4.2 Implementation

For hyper-parameters, we set the word embedding size d_w is 300. The sizes of part-of-speech embedding and named entities embedding are 20 and 10, respectively. The dimensionality of LSTM hidden states is 150. The initial learning rate is 0.001. The batch size is 32. We apply dropout on the input embeddings and outputs of LSTM, and the dropout rate is 0.4 and 0.5, respectively.

For trainable parameters, we initialize these parameters by randomly sampling from the uniform distribution in $[-0.1, 0.1]$. The word embeddings used in our experiment are pre-trained using GloVe model [17].

4.3 Compared with State-of-the-Art Methods

In Tabel 1, we compare our proposed model with previous state-of-the-art methods on the SemEval-2010 Task 8 dataset. The SVM model [19] is the top performed system. It achieves F_1-score of 82.2% by feeding a variety of hand-crafted features from many external resources. Socher et al. [22] build the recursive neural networks along the constituency tree for relation classification. They extend the recursive neural networks with matrix-vector spaces (MVRNN) [21] and improve the F_1-score to 82.4%. Zeng et al. [30] exploit the convolutional neural networks (CNN) to extract sentence level features and design position features for representing each word position in sentence. They concatenate the sentence

Table 1. Experimental results of our proposed model against other models.

Model	Feature set	F1
SVM [19]	POS, prefixes, morphological, WordNet, dependency parse, Levin classes, ProBank, FrameNet, NomLex-Plus, Google n-gram, paraphrases, TextRunner	82.2
RNN [22]	WE	74.8
	+POS, NER, WordNet	77.6
MVRNN [21]	WE	79.1
	+POS, NER, WordNet	82.4
CNN [30]	WE	69.7
	+position embeddings, words around nominals, WordNet	82.7
CR-CNN [20]	WE	82.8
	+position embeddings	84.1
SDP-LSTM [28]	WE	82.4
	Word, POS, GR, WordNet embeddings	83.7
DepNN [14]	WE, WordNet	83.0
	WE, NER	83.6
BLSTM [32]	WE	82.7
	+ Attention	84.0
Ours	WE	83.6
	+ PE, NE	**84.3**

level and lexical level features into a single vector for prediction. dos Santos et al. [20] propose CR-CNN model by replacing softmax loss function with ranking loss function. By omitting the impact of the "Other" class, they evaluate the F_1-score to 84.1%. Xu et al. [28] design SDP-LSTM model by building multiple LSTMs along the shortest dependency paths and achieve F_1-score of 83.7%. Zhou et al. [32] leverage bidirectional LSTM to extract sentence level features and exploit attention mechanism to obtain important information. Their proposed model achieve an F_1-score of 84.0%.

We can observe that our model achieves the competitive performance compared with state-of-the-art methods without any prior NLP knowledge. One of the reason is that the adversarial training improves the robustness and generalization of our proposed model. The another reason is that our proposed attention based gate mechanism is able to better choose the more discriminative features and concatenate the sentence level and lexical level features. When using word embeddings, part-of-speech embeddings and named entities embeddings as input, our proposed model yields a better performance than previous state-of-the-art methods. The result verifies that the part-of-speech embeddings and named entities embeddings contain rich semantic and syntactic information.

Table 2. Experimental results of the main model and its simplified models.

Model	Feature set	F1
BLSTM	WE	79.9
	+PE	80.6
	+NE	81.1
BLSTM+ATT	WE	81.0
	+PE	81.8
	+NE	82.1
BLSTM+AT	WE	81.0
	+PE	81.3
	+NE	81.6
BLSTM+GATE	WE	80.6
	+PE	81.1
	+NE	81.7
BLSTM+AT+GATE	WE	81.4
	+PE	81.6
	+NE	82.0
BLSTM+ATT+AT	WE	82.9
	+PE	83.2
	+NE	83.4
BLSTM+ATT+AT+GATE(Main)	WE	83.6
	+PE	84.1
	+NE	84.3

4.4 Effectiveness of Adversarial Training and Attention Based Gate Mechanism

Table 2 provides the experimental results of our proposed model (Main) and several simplified models. Several simplified models are described as follows:

- BLSTM: The model is used as strong baseline in our work, without adversarial training and attention based gate mechanism. We directly concatenate sentence level and lexical level features for prediction.
- BLSTM+ATT: We introduce attention mechanism based on BLSTM model to learn the important information.
- BLSTM+AT: Compared with BLSTM model, the BLSTM+AT model incorporates adversarial training to improve robustness and generalization performance.
- BLSTM+GATE: We apply gate mechanism into BLSTM model for adaptively combining sentence level and lexical level features.
- BLSTM+AT+GATE: Base on BLSTM model, we introduce adversarial training and gate mechanism.
- BLSTM+ATT+AT: Compared with baseline, the BLSTM+ATT+AT model integrates attention and adversarial training.

From the experimental results of Table 2, we have following observations:

- **Effectiveness of Attention Mechanism.** BLSTM+ATT model improves F1 score from 79.9% to 81.0% without linguistic knowledge, which demonstrates that attention mechanism is helpful for relation classification.
- **Effectiveness of Adversarial Training.** By introducing adversarial training, BLSTM+AT boosts the performance as compared with BLSTM model, showing 1.1% improvement. It proves that adversarial training is very effective for relation classification.
- **Effectiveness of Gate Mechanism.** When compared with BLSTM model, BLSTM+GATE improves the performance with the help of dynamically combining sentence level and lexical level features, which indicates the effectiveness of gate mechanism.

Our main model significantly outperforms other simplified models on this dataset. When adversarial training and attention based gate mechanism are combined (denoted as BLSTM+ATT+AT+GATE), we achieve the best F_1-score of 84.3%.

4.5 Case Study and Error Analysis

Further, we analyze some misclassification examples produced by our proposed model. Some typical wrongly classified instances are listed as follows:

Sample 1: "Then, the target PET $<e_1>$bottle$</e_1>$ was put inside of a metal $<e_2>$container$</e_2>$, which was grounded." with label Entity-Destination(e_1,e_2). However, this sentence is wrongly classified into Content-Container(e_1,e_2) by our model. To a certain extent, there is "containing" meaning between "bottle" and "container", so our model classifies the two entities into "Content-Container" relation. In addition, the "container" word exists in the sentence, which confuses the classifier to make a wrong prediction.

Sample 2: "People now post their $<e_1>$opinions$</e_1>$ to this $<e_2>$blog $</e_2>$." This sentence is classified into Entity-Destination(e_1,e_2) class, while the ground-truth label is "Other". The "opinions" can be seen as a entity, thus, there can be Entity-Destination(e_1,e_2) relation between the "opinion" and "blog".

Based on the analysis of misclassifications, more informative embeddings and higher level semantic features are required for making correct predictions.

5   Conclusion

In this paper, we exploit a bidirectional long short term memory network to extract sentence level features. In order to enhance the robustness and generalization of the proposed model, we apply adversarial training into BLSTM model. Additionally, we leverage attention mechanism to better learn the most influential features with respect to the two entities of interest. Further, to sufficiently exploit different level features, we dynamically concatenate the sentence level and lexical level features via gate mechanism. Experimental results on the SemEval-2010 Task 8 dataset show that our proposed model achieves better performance than previous state-of-the-art approaches.

Acknowledgments. This work is supported by the Natural Science Foundation of China (No. 61533018, No. 61702512 and No. 61502493). This work was also supported by Alibaba Group through Alibaba Innovative Research (AIR) Program and Huawei Tech. Ltm through Huawei Innovation Research Program.

References

1. Bahdanau, D., Cho, K., Bengio, Y.: Neural machine translation by jointly learning to align and translate. arXiv preprint arXiv:1409.0473 (2014)
2. Bunescu, R.C., Mooney, R.J.: A shortest path dependency kernel for relation extraction. In: Proceedings of the Conference on Human Language Technology and Empirical Methods in Natural Language Processing, pp. 724–731. ACL (2005)
3. Chen, J., Tandon, N., de Melo, G.: Neural word representations from large-scale commonsense knowledge. In: 2015 IEEE/WIC/ACM International Conference on Web Intelligence and Intelligent Agent Technology (WI-IAT), pp. 225–228 (2015)
4. Chen, J., Ji, D., Tan, C.L., Niu, Z.: Unsupervised feature selection for relation extraction. In: Companion Volume to the Proceedings of Conference Including Posters/Demos and Tutorial Abstracts (2005)
5. Elman, J.L.: Finding structure in time. Cogn. Sci. **14**(2), 179–211 (1990)
6. Goodfellow, I., et al.: Generative adversarial nets. In: Advances in Neural Information Processing Systems, pp. 2672–2680 (2014)
7. Goodfellow, I.J., Shlens, J., Szegedy, C.: Explaining and harnessing adversarial examples. arXiv preprint arXiv:1412.6572 (2014)
8. GuoDong, Z., Jian, S., Jie, Z., Min, Z.: Exploring various knowledge in relation extraction. In: Proceedings of the 43rd Annual Meeting on Association for Computational Linguistics, pp. 427–434. Association for Computational Linguistics (2005)
9. Hashimoto, K., Miwa, M., Tsuruoka, Y., Chikayama, T.: Simple customization of recursive neural networks for semantic relation classification. In: Proceedings of the 2013 Conference on EMNLP, pp. 1372–1376 (2013)
10. Hendrickx, I., et al.: Semeval-2010 task 8: multi-way classification of semantic relations between pairs of nominals, In: Proceedings of the Workshop on Semantic Evaluations: Recent Achievements and Future Directions, pp. 94–99. Association for Computational Linguistics (2009)
11. Hermann, K.M., et al.: Teaching machines to read and comprehend. In: Advances in Neural Information Processing Systems, pp. 1693–1701 (2015)
12. Hochreiter, S., Schmidhuber, J.: Long short-term memory. Neural Comput. **9**(8), 1735–1780 (1997)
13. Kingma, D.P., Ba, J.: Adam: a method for stochastic optimization. arXiv preprint arXiv:1412.6980 (2014)
14. Liu, Y., Wei, F., Li, S., Ji, H., Zhou, M., Wang, H.: A dependency-based neural network for relation classification. arXiv preprint arXiv:1507.04646 (2015)
15. Miyato, T., Dai, A.M., Goodfellow, I.: Adversarial training methods for semi-supervised text classification. arXiv preprint arXiv:1605.07725 (2016)
16. Mooney, R.J., Bunescu, R.C.: Subsequence kernels for relation extraction. In: Advances in Neural Information Processing Systems, pp. 171–178 (2006)

17. Pennington, J., Socher, R., Manning, C.: Glove: global vectors for word representation. In: Proceedings of the 2014 Conference on Empirical Methods in Natural Language Processing (EMNLP), pp. 1532–1543 (2014)
18. Qian, L., Zhou, G., Kong, F., Zhu, Q., Qian, P.: Exploiting constituent dependencies for tree kernel-based semantic relation extraction. In: Proceedings of the 22nd International Conference on Computational Linguistics, pp. 697–704. ACL (2008)
19. Rink, B., Harabagiu, S.: UTD: classifying semantic relations by combining lexical and semantic resources. In: Proceedings of the 5th International Workshop on Semantic Evaluation, pp. 256–259. Association for Computational Linguistics (2010)
20. Santos, C.N.D., Xiang, B., Zhou, B.: Classifying relations by ranking with convolutional neural networks. arXiv preprint arXiv:1504.06580 (2015)
21. Socher, R., Huval, B., Manning, C.D., Ng, A.Y.: Semantic compositionality through recursive matrix-vector spaces. In: Proceedings of the 2012 Joint Conference on Empirical Methods in Natural Language Processing and Computational Natural Language Learning, pp. 1201–1211. Association for Computational Linguistics (2012)
22. Socher, R., Pennington, J., Huang, E.H., Ng, A.Y., Manning, C.D.: Semi-supervised recursive autoencoders for predicting sentiment distributions. In: Proceedings of the Conference on EMNLP, pp. 151–161. ACL (2011)
23. Szegedy, C., et al.: Intriguing properties of neural networks. arXiv preprint arXiv:1312.6199 (2013)
24. Wu, F., Weld, D.S.: Open information extraction using Wikipedia. In: Proceedings of the 48th Annual Meeting of the Association for Computational Linguistics, pp. 118–127. Association for Computational Linguistics (2010)
25. Wu, Y., Bamman, D., Russell, S.: Adversarial training for relation extraction. In: Proceedings of the 2017 Conference on Empirical Methods in Natural Language Processing, pp. 1778–1783 (2017)
26. Xie, Z., et al.: Data noising as smoothing in neural network language models. arXiv preprint arXiv:1703.02573 (2017)
27. Xu, K., et al.: Show, attend and tell: neural image caption generation with visual attention. In: International Conference on Machine Learning, pp. 2048–2057 (2015)
28. Xu, Y., Mou, L., Li, G., Chen, Y., Peng, H., Jin, Z.: Classifying relations via long short term memory networks along shortest dependency paths. In: Proceedings of the 2015 Conference on Empirical Methods in Natural Language Processing, pp. 1785–1794 (2015)
29. Yao, X., Van Durme, B.: Information extraction over structured data: question answering with freebase. In: Proceedings of the 52nd Annual Meeting of the Association for Computational Linguistics, pp. 956–966 (2014)
30. Zeng, D., Liu, K., Lai, S., Zhou, G., Zhao, J.: Relation classification via convolutional deep neural network. In: Proceedings of COLING 2014, the 25th International Conference on Computational Linguistics: Technical Papers, pp. 2335–2344 (2014)
31. Zhang, S., Zheng, D., Hu, X., Yang, M.: Bidirectional long short-term memory networks for relation classification. In: Proceedings of the 29th Pacific Asia Conference on Language, Information and Computation, pp. 73–78 (2015)
32. Zhou, P., et al.: Attention-based bidirectional long short-term memory networks for relation classification. In: Proceedings of the 54th Annual Meeting of the Association for Computational Linguistics (Volume 2: Short Papers), vol. 2, pp. 207–212 (2016)

A Novel Approach on Entity Linking
for Encyclopedia Infoboxes

Xufeng Li[1][(✉)], Jianlei Yang[1], Richong Zhang[1], and Hongyuan Ma[2]

[1] BDBC and SKLSDE, School of Computer Science and Engineering,
Beihang University, Beijing 100190, China
{lixf17,jianlei,zhangrc}@act.buaa.edu.cn
[2] CNCERT/CC, Beijing 100020, China
mahongyuan@foxmail.com

Abstract. The Infoboxes in encyclopedia articles contain the structured factoid knowledge and have been the most important source for open domain knowledge base construction. However, if the hyperlink is missing in the Infobox, the semantic relatedness cannot be created. In this paper, we propose an effective model and summarize the most possible features for the infobox entity linking problem. Empirical studies confirm the superiority of our proposed model.

Keywords: Knowledge base · Information extraction · Entity linking

1 Introduction

Over the past few years, numerous knowledge base (KB) have been built and used in various fields, such as Question Answering, knowledge reasoning, knowledge representation and so on. Well-known KBs include Google Knowledge, DBpedia [1], Freebase, YAGO [15] and Zhishi.me [12]. To build a large scale domain-independent knowledge base, online encyclopedias, such as Wikipedia[1] and BaiduBaike[2], have become the important knowledge resources.

One critical reason that makes online encyclopedia extremely valuable is the human-edited structured data. Especially, the infobox consists of a set of attribute-value pairs summarizing the key aspects of a specific entity and this offers clean knowledge when building KB. The values in the infobox sometimes are linked to another page through hyperlinks. One can match the attribute of an infobox as a KB relation and the value corresponds to the attribute as a KB entity. These information is then transformed to RDF triples [1].

However, the hyperlink for the values of attributes in the infobox is not always available. This fact leads to a lacking of valuable relation associated with each entity. For example, in BaiduBaike, in the infobox of the entity *Scarlett Johansson*, the value for the attribute of notable work is "The Avengers", but

---

[1] www.wikipedia.com.
[2] baike.baidu.com.

© Springer Nature Singapore Pte Ltd. 2019
J. Zhao et al. (Eds.): CCKS 2018, CCIS 957, pp. 103–115, 2019.
https://doi.org/10.1007/978-981-13-3146-6_9

"The Avengers" appears just as a string and is not linked to the page of entity *The Avengers*. Therefore, it is necessary to predict these missing links to increase the connectivity in KB.

The ambiguity of entity name and the inconsistency of surface names make this task challenging. The same string may correspond to more than one entity and the same entity may have different names. Traditional algorithms solely extract some hand-craft features to characterize the relatedness between the infobox mention and the KB entities. However, the number of features is relatively small and many important features are not included. In addition, existing studies [19] only consider linear discriminative models, e.g. logistic regression, to predict the relatedness of the mention-entity pair. Usually, linear models can hardly model the practical complexity based on only a small number of features.

To solve these limitation of existing models, in this paper, we extend the feature set proposed in [19] and include more possible features to improve the performance of algorithms. In addition, we propose a boosting-based approach using GBDT [5] and logistic regression to increase the capacity of the classification model. Empirical studies confirms the effectiveness of the proposed model.

2   Related Work

2.1   Entity Linking

With the emergence of large KBs, entity linking from textual mentions to the KB entities has attracted research attentions. Several methods [6–8,14,16] have been proposed to tackle this problem. In general, three types of entity linking approaches have been exploited for mapping a textual entity mention to the corresponding real entity in KB.

Independent-Linking Methods: Early entity linking approaches believed that the similarity between the correct entity and the context of mention is higher than other candidate entities. So the result is linked to an entity which has the highest contextual similarity. For example, [10] merely adopted Bag of Words(BoW) model to do entity linking via calculating the cosine similarity between the context of the mention and the text of the candidate entity. In [2], extra knowledge is introduced such as category information in KB to overcome the weakness of BoW model and SVM is used to improve the performance of entity linking. In [7], a deep neural network method was used to learn the representation of an input document containing mention as well as documents referring to candidate entities.

Pair-Linking Methods: Pair-linking method is proposed to use the semantic association between candidate entities and the contextual entities, which can effectively enhance the performance of entity linking. Among numerous methods in this field, [11] proposes to utilize the semantic relatedness between candidate entities and other unambiguous entities with Normalized Google Distance (NGD) [18]. In [14], named entities were linked unifying Wikipedia and WordNet by leveraging the rich semantic knowledge embedded in Wikipedia and YAGO.

Graph-Based Methods: Considering the interdependence between entities in the same document, graph based approach is proposed to exploit the global interdependence of all entities for joint reasoning. In [6], authors model the global interdependence between different entity linking decisions. To collectively infer the referent entities of name mentions in the same document, they take the mention-entity context similarity and semantic relation between entities as the edge over a weighted Referent Graph, and reinforce the evidence according to the dependency structure between mention and the entities captured in Referent Graph until convergence.

2.2 Infobox Entity Linking

The above mentioned approaches only focus on mapping identified entities from text to KBs, but infobox information is also one of the most important source. To populate the missing hyperlinks in the Infobox, surface name matching and relatedness ranking are two major methods.

Surface Matching: In [12,17], the values of object properties are directly extracted from infoboxes and just name matching is used to populate missing links. In specific, if there exists an exact matched name with the property value, then the property value is replaced by the link to the matched entity. This method is simply a string match without considering entity linking techniques, thus the ambiguity problem cannot be properly solved.

Relatedness Ranking: In [19], authors first identified entity mentions in the Infobox, and then defined seven features for candidate entities. They exploit a logistic regression model to characterize the dependencies between candidate entity features and real unambiguous entities.

3 The Framework of Entity Linking

The goal of the entity linking is to predict the linked entities for infobox properties. Three subtasks are considered in our proposed framework. The first task focuses on identifying mentions in property values that may link to candidate entities; the second task aims at getting features of candidate entity-mention pair; and the third task aims at scoring all candidate entities by ranking module, then choosing the top-ranked entity as the target entity. In this study, we propose three main modules to solve the above mentioned subtasks.

3.1 Problem Formalization

In this section, the formal definition for entity linking over infoboxes will be introduced. For an entity e, the surface form of its property value refers to an entity is called a mention m, where m is associated with entity e through property name (relation) r. All candidate entities are denoted by $E(m)$. The goal of this research is to discover the true linked entity for mention m from $E(m)$.

Entity Page: Each entity page in encyclopedia describes a single entity and contains information on this entity. In most cases, the title of each page is the standard name for the entity described in this page. We extract the name of the entity as a mention, and the ID of entity as its value. It should be noted that, when the name of the entity is ambiguous in encyclopedia, the title has bracket to specify the ambiguous entity. For example, the article for entity *tennis player Li Na* has the title "Li Na (Chinese women's tennis star)", the bracket is removed to indicate the mention, i.e., "Li Na" in this example.

Disambiguation Page: Disambiguation pages is created to distinguish different entities with the same name. For example, the disambiguation page lists 47 different entities with the same name of "Li Na", including the famous tennis player, the pop singer and the others.

Alias in Infobox: The property of alias in infobox usually describes the alias information of the entity. It is very useful in extracting abbreviations and other aliases of entities. For example, the alias property in *Shaquille O'Neal* is "Big Shark", indicating that if "big shark" is mentioned, it may be linked to entity *Shaquille O'Neal*.

Anchor Text in Hyperlinks: Articles in encyclopedia often contain many anchor text which link to other entities. The anchor text, called hyperlink, provides the connection surface form and synonyms or variations of the linked entity. It is worth mentioning that when extracting the mention-entity pair from the anchor text, we also record the count of mention-entity pair at the same time. The count reflects the popularity of entities given the mention. For example, for the surface form "Li Na", the entity *Li Na (singer)* is much rare mentioned than *Li Na (player)*. In most cases when people mention "Li Na", they mean that *tennis player Li Na* rather than the *pop singer Li Na*.

3.2   Candidate Entities Generation Module

The candidate entities generation is mainly based on the string comparison of mentions and names of entities in the KB. To generate $E(m)$ which contains possible entities, it is necessary to build a mention-entity dictionary D that contains a vast amount of information on the surface forms of entities [13], such as name variations, abbreviations, aliases, etc.

The content of encyclopedia provides a set of useful features for generating candidate entities. Inspired by [14], we extract existing information in the encyclopedia to establish the mapping between the mention and the entity, including the article, the disambiguation page, the "alias" property in the infobox and the anchor text. We generate a dictionary D and record the times c pointing to the target entity. Each item in the dictionary $<mention, entity, count>$ represents the name of mention, its corresponding entity and the specific number of occurrences of the mention-entity pair. The number of occurrence reflects the popularity of the mentioned entity. The dictionary D is constructed as shown in Table 1 below.

Table 1. Mention-entity dictionary $\mathcal{D}$

Mention	Mapping entity (store the ID of entity actually)	Count
O'Neill	Shaquille O'Neal	155
Big Shark	Shaquille O'Neal	50
	Shark (fish in the ocean)	23
Shaquille O'Neal	Shaquille O'Neal	122
Li Na	Li Na (Chinese women's tennis star)	105
	Li Na (pop singer)	28
	Li Na (Qingdao movie actor)	21
Apple	Apple Inc	295
	Apple (Rosaceae apple fruit)	172
	Apple (Li Yu directed film in 2007)	14

3.3   Candidate Entity Ranking Module

Once the mention-entity dictionary $\mathcal{D}$ is built, we can identify the mention m that have not yet been linked and retrieve $E(m)$ from $\mathcal{D}$. However, in most cases, $|E(m)|$ is greater than one. We use several features to encode each candidate entity-mention pair and adopt a pairwise approach to rank the candidate entities. Our basic guiding point is that mention largely refers to topic-related entity. These features can be divided into three types: entity specific features, entity-mention features and relation-specific features.

To formulate the feature functions, we use T_e to denote a set of out-coming entities in the text description of the entity e. I_e denotes a set of entities which are already linked in the infobox of e. A_e denotes a set of entities in the abstract of e, where an abstract is usually the first paragraph of the descripion. E_r denotes a set of entities which also have the property r. Table 2 shows all features, and the definition of features will be mentioned below in detail.

1. **Entity Specific Features** including $f_1 \sim f_4$, only rely on the candidate entity itself, not related to the context where the mention appears. Entity popularity is one of features which can be computed by the prior probability of a candidate entity.

$$f_1(m, e_j) = \frac{count(m, e_j)}{count(m)} \tag{1}$$

   where $count(m, e)$ denotes the number of times that mention m occurred as the anchor text points to the entity e, $count(m)$ denotes the number of times that m appears in the encyclopedia.

2. **Mention Context features** including $f_5 \sim f_8$, measure the relatedness between candidate entities with the textual features of mentions. Specifically, we adopt the Wikipedia Link-based Measure(WLM) proposed by [18] to calculate the semantic relatedness between entities. WLM is modeled from the

Table 2. Features of candidate entities

Features	Full names	Definition
f_1	Entity popularity	The entity popularity of e_j
f_2	Co-occurrence	The number of co-occurrences of e_i and e_j
f_3	Existed	Whether e_j exists in the text of e_i
f_4	Existed reversely	Whether e_i exists in the text of e_j
f_5	Content relatedness	Average SR between e_j and each entity in T_{e_i}
f_6	Infobox relatedness	Average SR between e_j and each entity in I_{e_i}
f_7	Abstract relatedness	Average SR between e_j and each entity in A_{e_i}
f_8	Textual similarity	Textual similarity between e_j and e_i
f_9	Type matching	Type matching score of e_i and e_j
f_{10}	Relation relatedness	Average SR between e_j and each entity in E_{r_j}
f_{11}	Common property	Average number of common properties of e_j and each entity in E_{r_j}

Normalized Google Distance, which is based on encyclopedia hyperlink structure. Given two entities e_i and e_j, the semantic relatedness between them is defined as follows:

$$SR(e_i, e_j) = 1 - \frac{\log(max(|X_{e_i}|, |X_{e_j}|)) - \log(X_{e_i} \cap X_{e_j})}{\log(|W|) - \log(min(|X_{e_i}|, |X_{e_j}|))} \quad (2)$$

where X_{e_i} and X_{e_j} are the set of entities that links to e_i and e_j respectively, and W is the set of all entities in encyclopedia. From the above equation we can observe that the more common incoming links two entities have, the higher the semantic relatedness is.

Then, the Content relatedness between e_i and e_j can be given by:

$$f_5(e_i, e_j) = \frac{1}{|T_{e_i}|} \sum_{e_k \in T_{e_i}} SR(e_k, e_j) \quad (3)$$

Note that when computing the Infobox relatedness f_6, Abstract relatedness f_7 and Relation relatedness f_{10}, we can simply substitute the set of T_{e_i} by I_{e_i}, A_{e_i} and E_{r_j} respectively. In addition, the textual similarity is used to compare the description of e_i with the description of e_j. We generate Bag of words vectors with TF-IDF as their representation, and measure the cosine similarity, where V_{e_i} and V_{e_j} are represented as the textual vector of e_i and e_j.

$$f_8(e_i, e_j) = \frac{V_{e_i} \cdot V_{e_j}}{||V_{e_i}|| \times ||V_{e_j}||} \quad (4)$$

3. **Relation Specific Features** include $f_9 \sim f_{11}$. We take into consideration the relation-related features to make use of the entity type at both ends of a particular relation. These features can be used to estimate what type of entities

are more likely to be linked by the concerned relation. For example, in a triple <subject type, relation, object type>, given a relation "Representative works", when the subject type is "Singer", the linked entity type is likely to be "Song". When subject type is "Actor", then the linked entity type is more likely to be "Film and television works". Therefore we can first regard the category information in the encyclopedia as the entity type. Then, for each relation, we enumerate all entity types that associated with this relation and record the number of times. Table 3 shows an example of relation "Representative works" and its subject/object types.

After collecting the entity types and relations, the type matching score of each candidate entity can be given.

$$f_9(e_i, e_j) = \sum_{st \in Type(e_i)} \frac{1}{count(st, r_j)} \sum_{ot \in Type(e_j)} count(st, r_j, ot) \tag{5}$$

Table 3. The subject/object type and occurrence times of relation "Representative works"

Subject type	Object type
Singer	[["Music works", 1375], ["Film and television works", 1055], ["Entertainment works", 796], ["TV play", 783], ["Song", 673], ["TV drama works", 621], ["Music", 530], ["Film", 501], ["Pop music", 401], ["Album", 301]]
Actor	[["Film and television works", 5054], ["TV play", 3144], ["Film", 2898], ["TV drama works", 2547], ["Entertainment works", 1589], ["Entertainment", 873], ["Film works", 854], ["Dramas", 771], ["Dramas film", 523]]

Ranking Model. We can generate a feature vector $f(m, e) = (f_1, f_2, \ldots, f_{11})$ by concatenating above mentioned feature functions for each mention-entity pair of mention m and candidate entity e. Then the disambiguation problem is translated to the problem of defining a score function $s \circ f(m, e)$.

Given a mention m and its corresponding candidate entity set $E(m)$, the ranking model should score the correct entity higher than the rest candidate entities of the mention. The focus is on how to estimate the corresponding feature weights. So in prediction process, we calculate the score between a mention and each candidate entity and then select the entity with the highest score as the final result.

To effectively training the model, let $e_i > e_j$ denotes that entity e_i should be ranked higher than e_j. That is to say, the entity e_i has been labeled as "true target entity". We use the pairwise model to denote the ranking order between candidate entities. First, for each mention m, we get the corresponding entity e_i

and the other candidate entities e_j. Then we can set $<f(m, e_i) - f(m, e_j), 1>$ as positive example and $<f(m, e_j) - f(m, e_i), -1>$ as negative example, where $f(m, e_i)$ and $f(m, e_j)$ denote the feature vector of e_i and e_j respectively. In this way, we convert a ranking problem into a classification problem.

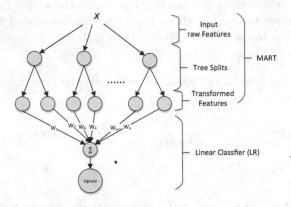

Fig. 1. The ranking model of GBDT and LR

As Fig. 1 shows, the gradient boosting decision tree [5] (GBDT or MART) and logistic regression (LR) model are exploited to learn the ranking model. The advantage of GBDT is that it is powerful to transform features to improve the performance of linear classifiers. Specifically, the core idea of GBDT is that in each iteration, a new decision tree is produced to model the gradient of the loss function with respect to the current model score evaluated at each training point. After several iterations of tree generation, we construct several weak tree model along the gradient direction and combine them with certain weights to form a strong model of final decision. After getting the final ensembled model, given a sample, it will fall on one leaf of each tree, and generate a binary vector based on whether it falls on each leaf. A binary vector is input to the linear classifier as a transformed vector. LR is then used to characterize the probability of e_i being ranked higher than e_j via a sigmoid function:

$$f(x) = P(e_i > e_j) = \frac{1}{1 + e^{-w(f(m,e_i) - f(m,e_j))}} \tag{6}$$

In GBDT, we use the exponential loss function as follows:

$$L(y, f(x)) = \log(1 + e^{-yf(x)}) \tag{7}$$

where $y \in \{-1, +1\}$, denotes the true label.

The model parameters can be estimated by taking derivative of the loss with respect to the whole set of parameters as in Algorithm 1. Line 1 to 8 is training a GBDT tree model, where R_{mj} denotes the set of data samples that land in the

$j^{th}(j = 1, 2, \ldots, J)$ leaf node of the mth tree and each new tree are generated to fit r_{mi}, the optimal fitting value of each leaf is c_{mj}. After getting the final ensembled tree and transforming vector based on whether sample falls on each leaf in Line 10, we can get new transformed features and labels T'. As Line 13 shows, cross entropy is used as loss function and weights are estimated through gradient descent method, where α is the learning rate.

Algorithm 1. Training MART and LR model for estimating weights

---

Input: Training samples $T = \{(x_1, y_1), (x_2, y_2), \ldots, (x_N, y_N)\}, x_i \in R^n, y_i \in \{-1, +1\}$;

Output: The weights vector $W = (w_1, w_2, , w_n)$

1: Initialize $f_0(x) = \arg\min_c \sum_{i=1}^{N} L(y_i, c)$

2: **for** *every iteration* $m = 0$ to M **do**

3:    (a) For every sample $i = 1, 2, \ldots, N$, compute the negative gradient error of the loss function is: $r_{mi} = -\left[\frac{\partial L(y_i, c)}{\partial f(x_i)}\right]_{f(x)=f_{m-1}(x)} = \frac{y_i}{(1+e^{y_i(f_{m-1}(x)+c)})}$

4:    (b) For $j = 1, 2, \ldots, J$, the optimal fitting value of each leaf is: $c_{mj} = \arg\min_c \sum_{x_i \in R_{mj}} \log(1 + e^{y_i(f_{m-1}(x_i))+c})$

5:    (c) Update new tree: $f_m(x) = f_{m-1}(x) + \sum_{j=1}^{J} c_{mj} I(x \in R_{mj})$

6: **end for**

7: Get the final ensembled tree: $\hat{f}(x) = f_M(x) = \sum_{m=1}^{M} \sum_{j=1}^{J} c_{mj} I(x \in R_{mj})$

8: **for** *Every sample* $n = 1$ to N **do**

9:    $\hat{f}(x_i) \xleftarrow{\hat{f}(x)} x_i$

10: **end for**

11: $T' = \{(\hat{f}(x_1), y_1), (\hat{f}(x_2), y_2), \ldots, (\hat{f}(x_N), y_N)\}$

12: Minimize: $L(w) = \sum_{i=1}^{|N|} -y_i \ln \frac{1}{1+e^{-w\hat{f}(x_i)}} - (1 - y_i) \ln (1 - \frac{1}{1+e^{-w\hat{f}(x_i)}})$

13: Using gradient descent: $w = w + \alpha \frac{\partial L(w)}{\partial w}$

---

4 Experiments and Evaluation

4.1 Dataset

To evaluate the performance of the proposed approach, we crawl articles from BaiduBaike which contain information including infoboxes, description that provides a concise summary of salient information of entity with hyperlink, categories, etc. We remove some non-standard characters, separate for multi-attribute values over infoboxes, and remove a small number of entities that category unassociated. We finally collect 9,568,051 entity articles, 31,578 relations and 1,722 entity types.

4.2   Training Data Generation

It is commonly accepted that large scale training data is crucial for training rank-ing model. In order to obtain massive amount of training data without tremen-dous manual annotation efforts, we collect anchor texts that contain hyperlink to the reference entity from infoboxes. Suppose for an entity e, its infobox has several attributes, one of attribute value m is an anchor text, and it is linked to an entity e_m. And get the rest candidate entities for m in the mention-entity dic-tionary $\mathcal{D}$ to construct positive and negative examples according to the Sect. 3.3 mentioned.

We randomly choose 10,669 mention-entity pairs, in which every mention m has a true corresponding entity e_m and a few other candidate entities. As above said, each candidate entity constitute a pairwise model with the true entity e_m to generate a positive example and a negative example. Therefore each mention-entity pair could generate several training examples. In our experiment, 45,709 positive and 45,709 negative examples are generated to evaluate the effectiveness of our proposed model. Then, the data is randomly divided to 70% as training set and 30% test set respectively.

4.3   Baseline Model

We evaluate the performance of our proposed method as well as baseline methods for entity liking on dataset mentioned above. The details of other baseline models are described as follows.

BoW Model (BoW): In [3], this straightforward method is to compare the context of the mention and the description of candidate entities. Here, first, bags of words vectors is generated for the text description of the entities where mentions in and the text description of candidate entities. Then, the similarity is measured between the vectors. The candidate entity achieving the highest similarity score is chosen as the mapping entity.

LR Ranking Model (LR): In [19], the ranking problem is converted into a classification problem and solved with logistic regression. This method mini-mizes the cross entropy loss function to estimate parameters, which penalizes the deviation of the model output probabilities from the true probabilities to fit the loss.

SVM Ranking Model (SVM): In [4,14,16], a max-margin loss is used for training the model. This method supposes the output score of a correct entity should be higher than that of a randomly selected candidate entity by a margin of m. SVM-linear ranking model is proposed by [9] and the max-margin loss function is used.

4.4   Experimental Results

For each testing data, we choose the entity with highest score and use the average precision as the metric to evaluate the performance of compared approaches. All

Table 4. Comparison results of our proposed method and other baseline methods

Methods	Precision
BoW [3]	0.4931
LR [19]	0.8692
SVM [4,14,16]	0.8935
Our method	**0.9295**

Table 5. Different features evaluation results

Features	Precision
Entity Popularity (f_1)	0.8431
+($f_3, f_5 - f_7, f_{10}$) [19]	0.8669
+co-occurrence (f_2)	0.8769
+textual similarity (f_8)	0.8879
+type matching (f_9)	0.9192
Overall($f_1 - f_{11}$)	**0.9295**

hyper-parameters are tuned using 10-fold cross validation over the training data. Table 4 shows the experimental results of our proposed method and the other compared methods. As can be observed from the results, our method achieves the best performance.

4.5 Features Study

We conducted features analysis on our method to verify the effectiveness of incorporating different feature functions and the results are reported in Table 5. We begin with the most simple feature function, entity popularity. Then various features are incorporated into the model incrementally. It can be seen that every feature has a positive effect on the experimental results. When incorporating all the proposed feature functions proposed in this paper, we can achieve the best result. Surprisingly, the performance is significantly improved after adding the feature *type matching*, which indicats that entity type at both ends of the relation is useful to deal with entity linking over infoboxes.

5   Conclusion

This paper studies a problem of linking entities from the infobox to existing KB entities. As the disambiguation problem of entity linking widely exists, we proposed a boosting-based approach to transform the KB features into a latent vector and exploit a logistic regression model for the link prediction. A real-world dataset is collected and empirical studies confirms the effectiveness of our proposed method.

Acknowledgments. This work is supported partly by China 973 program (No.2015CB358700), by the National Natural Science Foundation of China (No. 61772059, 61421003). This paper is also supported by the State Key Laboratory of Software Development Environment of China and Beijing Advanced Innovation Center for Big Data and Brain Computing.

References

1. Auer, S., Bizer, C., Kobilarov, G., Lehmann, J., Cyganiak, R., Ives, Z.: DBpedia: a nucleus for a web of open data. In: Aberer, K., et al. (eds.) ASWC/ISWC -2007. LNCS, vol. 4825, pp. 722–735. Springer, Heidelberg (2007). https://doi.org/10.1007/978-3-540-76298-0_52
2. Bunescu, R.C., Pasca, M.: Using encyclopedic knowledge for named entity disambiguation. In: EACL, vol. 6, pp. 9–16 (2006)
3. Chen, Z., et al.: CUNY-BLENDER TAC-KBP2010 entity linking and slot filling system description. In: Theory and Applications of Categories (2010)
4. Dredze, M., McNamee, P., Rao, D., Gerber, A., Finin, T.: Entity disambiguation for knowledge base population. In: Proceedings of the 23rd International Conference on Computational Linguistics, pp. 277–285. Association for Computational Linguistics (2010)
5. Friedman, J.H.: Greedy function approximation: a gradient boosting machine. Ann. Stat. **29**(5), 1189–1232 (2001)
6. Han, X., Sun, L., Zhao, J.: Collective entity linking in web text: a graph-based method. In: Proceedings of the 34th International ACM SIGIR Conference on Research and Development in Information Retrieval, pp. 765–774. ACM (2011)
7. He, Z., Liu, S., Li, M., Zhou, M., Zhang, L., Wang, H.: Learning entity representation for entity disambiguation. In: Proceedings of the 51st Annual Meeting of the Association for Computational Linguistics, ACL 2013, pp. 30–34 (2013)
8. Huang, H., Heck, L., Ji, H.: Leveraging deep neural networks and knowledge graphs for entity disambiguation. arXiv preprint arXiv:1504.07678 (2015)
9. Joachims, T.: Training linear SVMs in linear time. In: Proceedings of the 12th ACM SIGKDD International Conference on Knowledge Discovery and Data Mining, pp. 217–226. ACM (2006)
10. Mihalcea, R., Csomai, A.: Wikify!: linking documents to encyclopedic knowledge. In: Proceedings of the Sixteenth ACM Conference on Information and Knowledge Management, pp. 233–242. ACM (2007)
11. Milne, D., Witten, I.H.: Learning to link with Wikipedia. In: Proceedings of the 17th ACM Conference on Information and Knowledge Management, pp. 509–518. ACM (2008)
12. Niu, X., Sun, X., Wang, H., Rong, S., Qi, G., Yu, Y.: Zhishi.me - weaving Chinese linking open data. In: Aroyo, L., et al. (eds.) ISWC 2011. LNCS, vol. 7032, pp. 205–220. Springer, Heidelberg (2011). https://doi.org/10.1007/978-3-642-25093-4_14
13. Shen, W., Wang, J., Han, J.: Entity linking with a knowledge base: issues, techniques, and solutions. IEEE Trans. Knowl. Data Eng. **27**(2), 443–460 (2015)
14. Shen, W., Wang, J., Luo, P., Wang, M.: Linden: linking named entities with knowledge base via semantic knowledge. In: International World Wide Web Conferences, pp. 449–458 (2012)
15. Suchanek, F.M., Kasneci, G., Weikum, G.: Yago: a core of semantic knowledge unifying WordNet and Wikipedia. In: International World Wide Web Conferences (2007)
16. Sun, Y., Lin, L., Tang, D., Yang, N., Ji, Z., Wang, X.: Modeling mention, context and entity with neural networks for entity disambiguation. In: Proceedings of the Twenty-Fourth International Joint Conference on Artificial Intelligence, IJCAI 2015, Buenos Aires, Argentina, 25–31 July 2015, pp. 1333–1339 (2015)

17. Wang, Z., Wang, Z., Li, J., Pan, J.Z.: Knowledge extraction from chinese Wiki encyclopedias. J. Zhejiang Univ. Sci. C **13**(4), 268–280 (2012)
18. Witten, I., Milne, D.: An effective, low-cost measure of semantic relatedness obtained from Wikipedia links. In: Proceeding of AAAI Workshop on Wikipedia and Artificial Intelligence: An Evolving Synergy, pp. 25–30. AAAI Press, Chicago (2008)
19. Xu, M., et al.: Discovering missing semantic relations between entities in Wikipedia. In: Alani, H., et al. (eds.) ISWC 2013. LNCS, vol. 8218, pp. 673–686. Springer, Heidelberg (2013). https://doi.org/10.1007/978-3-642-41335-3_42

Predicting Concept-Based Research Trends with Rhetorical Framing

Jifan Yu[1]([✉]), Liangming Pan[2], Juanzi Li[1], and Xiaoping Du[3]

[1] Tsinghua University, Beijing, China
{yujf18.mail,lijuanzi}@tsinghua.edu.cn
[2] National University of Singapore, Singapore, Singapore
e0272310@u.nus.edu
[3] Beihang University, Beijing, China
xpdu@buaa.edu.cn

Abstract. Applying data mining techniques to help researchers discover, understand, and predict research trends is a highly beneficial but challenging task. The existing researches mainly use topics extracted from literatures as objects to build predicting model. To get more accurate results, we use concepts instead of topics constructing a model to predict their rise and fall trends, considering the rhetorical characteristics of them. The experimental results based on ACL1965-2017 literature dataset show the clues of the scientific trends can be found in the rhetorical distribution of concepts. After adding the relevant concepts' information, the predict model's accuracy rate can be significantly improved, compared to the prior topic-based algorithm.

Keywords: Scientific trends analysis · Concept extraction
Scientific discourse analysis

1 Introduction

With the rapid growth of research community, it becomes increasingly difficult for researchers to see the complete picture of how a research field has been evolving, with the overwhelming amount of scientific literatures. Therefore, applying data mining techniques to help researchers discover, understand, and predict research trends becomes a highly beneficial but challenging task.

Most of the existing methods for research trends analysis are topic-based [2,4,6]. They first extract research topics using topic modeling on a collection of scientific papers, and then study the rise and fall of each topic based on machine learning or statistical methods. Among these methods, Vinodkumar et al. [13]. predicts trends of a topic based on the changing of its rhetorical role. The rhetorical role of a topic is the purpose or role it plays in the paper: such as background, research objectives, methods, conclusions, etc. Rhetorical functions that topics take part in serve as strong clues of the topic evolution. For example, if a topic was often to be employed as a method in the past, but was mentioned a

© Springer Nature Singapore Pte Ltd. 2019
J. Zhao et al. (Eds.): CCKS 2018, CCIS 957, pp. 116–128, 2019.
https://doi.org/10.1007/978-981-13-3146-6_10

lot as background recently may signal an increase in its maturity and a possible decrease in its popularity.

However, topic-based methods can not provide the level of granularity needed to support an in-depth analysis of research dynamics in scientific literature, due to the following reasons. First, a topic is a word distribution, which requires further human interpretation to understand its meaning. Many word distributions are noisy and hard to interpret. Second, topics are often too coarse-grained. For example, the topic about SVM may contain the algorithm of SVM, the applications of SVM as well as other SVM-based methods.

In this paper, we propose to analyze the research trends of scientific documents from a different perspective, i.e., concept-based analysis. Concepts in scientific papers are key phrases that express the main idea of the paper, for example, problems (e.g., NER), techniques (e.g., SVM), domains (e.g., machine translation), datasets (e.g., Semeval 2010), and evaluation metrics (e.g, F1 Score). Detecting the rise and fall of concepts rather than topics can provide a more fine-grained view of research dynamics. We follow the idea of Vinodkumar et al. to investigate how the change of rhetorical functions influences the rise and fall of scientific concepts. However, performing concept-based analysis poses several unique challenges. First, concepts in scientific papers are hard to be identified because most of them are newly-proposed and domain-specific. Second, the significant variability of expressions makes a concept often has many identical mentions (e.g., SVM and support vector machine). If we regard each mention as a separate concept, it will raise the data sparsity issue. However, clustering identical concept mentions is tough, because lexical similarity is not reliable for the clustering, and traditional clustering algorithms are hard to generate tight enough clusters.

To address the above challenges, we introduce a novel algorithm that uses the rhetorical structural features of the concept to analyze the rise and fall of scientific concepts. Specifically, we first propose a method to extract the concepts from the scientific literature and rationally merge the synonyms of them to obtain the concept data set. Then, based on the work of Vinodkumar et al. [13]., we automatically annotate each sentence of the abstract with a rhetorical role. Finally, based on different rhetorical roles of the concept at different times, we analyze the evolution trajectory of the concepts. The entire study was conducted on the ACL1965-2017 dataset consisting of 36,929 actual papers. The experimental results show that the rhetorical functions of the concept and the changing trends of its related concepts have significant effects on the growth or decline of its popularity in the future.

Contributions: The three main contributions of our paper are: (1) we proposed a new method to extract concepts from scientific literature and merge the identical concept mentions; (2) we show that the rhetorical function distribution of a concept also reflects its temporal trajectory, and that it is predictive of whether the concept will eventually rise or fall in popularity; (3) we significantly improve the prediction accuracy of the existing model by considering related concepts.

2   Related Works

Our work is based on the work of keyphrase extraction and scientific trend analysis. Keyphrase extraction provides research candidates to form concepts, while scientific trend analysis with scientific discourse analysis provides research ideas for this work.

2.1   Keyphrase Extraction

Keyphrases are defined as a set of terms in a document that give a brief summary of its content for readers. Automatic keyphrase extraction is widely used in information retrieval and digital library [17,25]. Keyphrase extraction is also an essential step in various tasks of natural language processing such as document categorization, clustering and summarization. There are two principled approaches to extracting keyphrases: supervised and unsupervised. In the unsupervised approach, graph-based ranking methods are state-of-the-art [16,26]. These methods first build a word graph according to word co-occurrences within the document, and then use random walk techniques to measure word importance. After that, top ranked words are selected as keyphrases.

The supervised approach [21] regards keyphrase extraction as a classification task, in which a model is trained to determine whether a candidate phrase is a keyphrase. Our work chooses this approach and groups identical key phrases to build the concept dataset.

2.2   Scientific Trends Analysis

In literature metrology and scientometrics, there are a lot of researches on the trends of scientific research. Research methods can be broadly divided into two types, one focusing on the citation of the literature and one focusing on textual information. The former researchers often used topological methods to identify those emerging research topics in advance from the common reference clustering [8,11] or mutual reference networks of the literature [7]. The other part starts with the text of the paper itself. For example, Mane and Guo use the word burst to find new and emerging scientific fields [9,10], while Small makes emotional assessments of the various cited texts and shows the different potential of scientific terms [12].

Vinodkumar's work [13] analyzes traditional scientific trend using scientific discourse analysis [24]. By dividing the literature abstracts into rhetorical functions, statistics on the frequency of occurrence of each scientific topic in different regions are analyzed, and finally contribute to predicting the trend of it. Our work gain the idea from it and optimize the model by changing research objects to concepts.

3   Methods

To provide an end-to-end solution of predicting the trends of scientific concepts, we design a two-stage framework. First, we automatically extract concept mentions from scientific papers, and then we propose a novel clustering algorithm to merge identical or similar concept mentions. Second, based on the work of Vinodkumar et al., we design a model that uses the rhetorical features to predict the rise and fall of scientific concepts. In the following sections, we will introduce the two parts in details.

3.1   Discovery of Scientific Concepts

Given a collection of scientific papers, we first extract concept mentions and design a novel algorithm to cluster identical concepts. The concept mentions are clustered in suitable granularity to address the data sparsity issue, and facilitate the trend prediction in the next stage.

Extracting Concept Mentions. We design a three-stage method to automatically extract scientific concepts from the literature as follows.

1. We first extract candidate concept mentions using linguistic patterns.
2. We then calculate the feature vector for each candidate mention.
3. Finally, we classify each candidate mention as valid scientific mention or not by training a binary classification model.

Step 1 Considering that most concepts are noun phrases [1], we obtain candidate course concepts by extracting all noun phrases in the paper using the following POS pattern, where JJ is presented as adjective, NN as noun and IN as preposition.

$$\{((\langle JJ \rangle^* \langle NN^* \rangle + \langle IN \rangle)? \langle JJ \rangle^* \langle NN^* \rangle +\} \tag{1}$$

Step 2 In order to filter out noise from the candidates, and obtain qualified concept mentions, we train a binary classifier using features using different aspects of information. Specifically, we calculate the feature vector for each candidate, and then train a binary classifier to determine whether a candidate is a valid concept mention. The feature design is shown in the Table 1.

Step 3 After feature engineering, we train a binary classifier to classify candidate as "concept mention" or "not concept mention".

Table 1. Feature engineering of noun-phrases.

Category	Feature	Description
Frequency-based features	Term frequency	The frequency of the term in the paper
	Sentence IDF	The percentage of sentences in which the term appears
	PMI	Point Mutual Information of the term in this paper
Statistical features	Term length	The length of the term
	Lexical cohesion	Lexical cohesion for term t is
	Max word length	The length of the longest word
Grammatical features	Is acronym	Whether the term is an acronym
	Is capital	Whether the first letter of this term is capital
	Is named entity	Whether the term is a named entity
Positional features	First occurrence	Normalized positions of first occurrence
	Last occurrence	Normalized positions of last occurrence
	Spread	Difference between first occurrence and last occurrence
	In title	Whether the term appears in the paper title
	In abstract	Whether the term appears in the paper abstract

Clustering Identical Concept Mentions. Unlike topics, the number of extracted concepts can be very big, which results in many infrequent concepts. Performing trend analysis on these infrequent concepts will raise the data sparsity problem. Therefore, after obtaining all concept mentions from the collection of scientific papers, we propose a mention clustering algorithm to automatically group identical and similar concept mentions. Different from traditional clustering algorithms (e.g., K-Means), our algorithm clusters similar concept mentions without specifying the cluster numbers, which is more suitable in our problem setting.

Before introducing the algorithm, we first propose two assumptions about the nature of identical concept mentions, based on observations on real-life scientific literature.

- **Co-occurrence**: Identical mentions tend to co-occur within close context windows.
- **Co-reference**: Different mentions of a same concept are likely to cite the same paper.

Figure 1 uses the concept "Word Embedding" as an example to show these two properties of identical mentions in actual texts. Authors use the abbreviation and synonym like "WEs", "vector representations of words", etc. to emphasize and explain this concept, and these all cite a same milestone paper.

Dense real-valued distributed representations of words known as **word embeddings (WEs)** (*Mikolov et al., 2013*) have become ubiquitous in NLP...

Recently, the **distributed representations of words** (*Mikolov et al., 2013*) have been shown that ...

... optimizes **vector representations of words (word embeddings)** (*Mikolov et al., 2013*) such that they can predict other context words occurring in a small window...

...and measuring the similarity with **word2vector** (*Mikolov et al., 2013*).

Fig. 1. Identical mentions often co-occur as explanations and cite same papers.

Based on the above assumptions, we could identify identical or similar concepts by capturing these two aspects of information, i.e., co-occurrence and co-reference information.

– We capture co-occurrence information by learning **concept embeddings** [23]. Word Embeddings aims to maximum the probability of the context words given the center word in a small context window, so we first replace each concept mention as a single token in the corpus, then train word-embeddings on it to obtain the semantic representations for each concept mention. The cosine distance between two mention vectors, denoted as $sim\,(m_i, m_j)$, can represent their semantic relatedness.
– We capture co-reference information by extracting **citing sentences**. We define a citing sentence as the sentence which contains at least one citation. For a citing sentence containing a concept mention m and cites a paper p, we obtain a mention-paper pair (m, p) and finally get the set of cited papers $cit\,(m)$ for each mention m.

Based on the above components, we propose the **Concept Mention Grouping Algorithm**. The algorithm iteratively merges similar concepts, and makes sure the following property holds throughout all the iterations. For concept mentions in the same cluster, the semantic relatedness between any two mentions must be larger than σ, and all of the mentions in the cluster must at least co-reference θ papers. The details of the algorithm is shown in Algorithm 1.

For a cluster c, we define $cit\,(c) = \bigcap_{m \in c} cit\,(m)$. We also define *cluster similarity* as $sim\,(c_i, c_j) = \min\{sim\,(m_1, m_2)\}$, when $m_1 \in c_i$ and $m_2 \in c_j$.

It can be proven that for any cluster c_i that Algorithm 1 outputs, we have $\forall m_h, m_l \in c_i, sim\,(m_h, m_l) \geq \sigma$ and $\left|\bigcap_{m \in c_i} cit\,(m)\right| \geq \theta$. This ensures that concept mentions with high co-occurrences and co-references are clustered together. As for the time complexity, Step 2 is of $O\left(n^2 log\,(n)\right)$, Step 3-11 is of $O\left(n^2\right)$, so the total time complexity is $O\left(n^2 log\,(n)\right)$.

Algorithm 1. Concept Mention Grouping Algorithm.

Input: Concept mentions as $M = \{m_1, \cdots, m_n\}$ and Cited Papers for each mention as $Cit(M) = \{cit(m_1), \cdots, cit(m_n)\}$.

Output: Concepts as $C = \{c_1, \cdots, c_k\}$, where $c_i = \left\{m_{i_1}, \cdots, m_{i_{|c_i|}}\right\}$, $c_1 \bigcup \cdots \bigcup c_k = M$ and $c_i \bigcap c_j = \emptyset$, $\forall c_i, c_j \in C$.

1: Initialize concepts $C = \{c_1, \cdots, c_n\}$, $c_i = \{m_i\}$;

2: Ranking by $sim(m_i, m_j)$ to obtain a list of tuples, where $\forall k$, $i_k \neq j_k$, $\forall k_1 < k_2$, $sim\left(m_{i_{k_1}}, m_{i_{k_1}}\right) \geq sim\left(m_{i_{k_2}}, m_{i_{k_2}}\right)$, $L = \left\{(i_1, j_1, sim(m_{i_1}, m_{j_1})), \cdots, \left(i_{n^2-n}, j_{n^2-n}, sim(m_{i_{n^2-n}}, m_{j_{n^2-n}})\right)\right\}$;

3: Pop the first tuple of L. $(i, j, sim(m_i, m_j)) = Pop(L)$;

4: **if** $sim(m_i, m_j) < \delta$ **then return** C, end

5: **end if**

6: Find m_i and m_j currently belong to which concept, denoted as c_{m_i} and c_{m_j}.;

7: Deleting some weak classifiers in E_n so as to keep the capacity of E_n;

8: **if** $sim\left(c_{m_i}, c_{m_j}\right) == sim(m_i, m_j)$, $cit(c_{m_i}) \bigcap cit(c_{m_j}) \geq \theta$ **then**

9:　　$C = Merge\left(c_{m_i}, c_{m_j}\right)$

10: **else**

11:　　GOTO **step 3**

12: **end if**

3.2　Predicting the Rise and Fall of Scientific Concepts

Rhetorical Feature Extraction and Modeling. To setup the rhetorical analysis, we first use a tool based on *CRF* named *ArgZoneTagger* [13] to place each sentence in abstract with seven different tags, which expresses its rhetorical attributes.

The seven tags are: BACKGROUND(The scientific context), OBJECTIVE(The specific goal), DATA(The empirical dimension used), DESIGN(The experimental setup), METHOD(Means used to achieve the goal), RESULT(What was found) and CONCLUSION(What was inferred). *LDP* are seven features corresponding to the percentage of concepts across the seven rhetorical function labels (e.g., % of time the concept is a METHOD like Fig. 2), which can be calculated as below. $LD(c, s_i, t)$ is the frequentness Concept c occurs in segment s_i during time t.

$$LDP(c, s_i, t) = LD(c, s_i, t) / \sum LD(c, s, t) \tag{2}$$

Digging further, we can also find other 7 features named *LDR*, reflecting how much a concept's *LDP* changed from former period to now.

$$LDR(c, s, t_i) = LDP(c, s, t_i) - LDP(c, s, t_{i-1}) \tag{3}$$

We also can combine *LDP* and *LDR* features to build a more completed 14-dimension feature named *LDS*, the experiment will test the behavior of these three features. Next, we simplified the task of predicting lifting into a classification problem. By calculating the difference between the heat in the next period

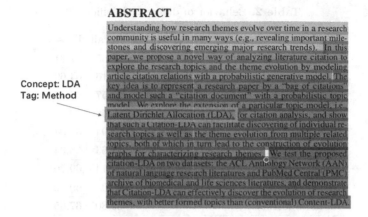

ABSTRACT

Understanding how research themes evolve over time in a research community is useful in many ways (e.g., revealing important milestones and discovering emerging major research trends). In this paper, we propose a novel way of analyzing literature citation to explore the research topics and the theme evolution by modeling article citation relations with a probabilistic generative model. The key idea is to represent a research paper by a "bag of citations" and model such a "citation document" with a probabilistic topic model. We explore the extension of a particular topic model, i.e. Latent Dirichlet Allocation (LDA), for citation analysis, and show that such a Citation-LDA can facilitate discovering of individual research topics as well as the theme evolution from multiple related topics, both of which in turn lead to the construction of evolution graphs for characterizing research themes. We test the proposed citation-LDA on two datasets: the ACL Anthology Network (AAN) of natural language research literatures and PubMed Central (PMC) archive of biomedical and life sciences literatures, and demonstrate that Citation-LDA can effectively discover the evolution of research themes, with better formed topics than (conventional) Content-LDA.

Concept: LDA
Tag: Method

Fig. 2. A tagged abstract and concept "LDA" occurs in "Method" segment.

of time and the current, we divide the concepts into concepts of ascent, concepts of decline, and concepts of stability. This work can be done by training a classifier based on L2 logistic regression.

Using Related Concepts to Optimize the Model. In fact, the lifting of a concept not only depends on itself. A rising concept may cause changes in its associated multiple concepts as topics do [20]. Therefore, it is also meaningful to investigate the trend of a few concepts related to a specific concept. Thanks to our discovering work, we can also use the similarity of two concepts as $sim(c_i, c_j)$ in Sect. 3.1 to find out the most related ones of concept c_i. Adding the recent trends of them as assistant features may also give a rise of the accuracy of our prediction.

4 Experiments

4.1 Results of Discovering Scientific Concepts

According to the process of finding the concept, we first extract all qualified noun phrases from the literature and conduct preliminary screening to obtain the following results, for Noun-phrases within 5-grams:

- Number of extracted candidate concept mentions: **21,308,601**
- Number of **unique** candidate concept mentions: **5,962,170**
- Number of candidate mentions **per paper: 577**

As what is mentioned in Sect. 3.1, we build a binary classifier to classify candidate as "concept mention" or "not concept mention" using Sem-Eval 2010 as our training data. It contains 144 ACL papers with human-annotated key phrases. For each paper, we use the feature vectors of human-annotated key

Table 2. Behavior of different classifiers.

Classifier	Precision	Recall	F1-score
K-nearest neighbors (K = 3)	70.12	65.65	67.81
L2 logistic	65.23	64.16	64.69
L1 logistic	75.51	62.57	68.43
Linear SVC	72.47	66.66	69.44
RBF SVM	72.44	65.94	69.04
Navie Bayesian	71.83	61.63	66.34
Decision tree	76.55	63.14	69.20
Ada boost	69.87	68.98	69.42
Random forest	77.87	59.80	67.65

phrases as positive examples, and randomly sample some candidate key phrases as negative examples. It results a total of 3683 training examples, with 1624 positive examples and 2059 negative examples. In order to select the most suitable classification algorithm, we tried almost all the mainstream algorithms and used 5-fold cross validation to evaluate the results of different classifiers. The results are displayed in detail in the Table 2.

In fact, the performance of mainstream algorithms is not much different, so we chose **Linear SVC** with the highest F1-Score as the classification algorithm, because it can also calculate the score of each phrase in the article to facilitate future work. Finally we get **13,002** unique concept mentions with three conditions:

1. The score given by the mention classifier must be **positive**.
2. For each paper, we only select its **top-20 scoring** candidate mentions.
2. The candidate must appears in **at least 5 papers**.

Since we have raised the **Concept Mention Grouping Algorithm** with the feature of citing reference, we first extract all citing sentences in the ACL Dataset resulting in below:

- Resulted in a **total of 538,956** citing sentences.
- With an **average of 14.59** citing sentences per paper.
- Number of **unique** cited papers is: **111,453**.

The thresholds of the semantic relatedness between any of a concept cluster's mentions, σ and the number of co-reference papers, θ can be adjusted to get different result of clustering work as Table 3. We select the better-grained case of $\theta = 1$, $\sigma = 0.7$ as a concept data set, and Table 4 shows a few examples in this case.

4.2   Results of Predicting the Rise and Fall of Concepts

Our task comes to be building a predicting model based on Concepts' rhetorical features. We first divide the ACL Dataset which contains 36,929 articles into 369

Table 3. Results of concept mention grouping algorithm

θ	σ	Number of clusters	Average mentions per cluster
3	0.6	8,326	1.56
3	0.7	9,136	1.42
1	0.6	1,730	7.51
1	0.7	4,557	2.85

Table 4. Examples of concept clusters($\theta = 1$, $\sigma = 0.7$)

Concept	Mentions	Concept	Mentions
5	metadata information meta-information meta information metadata	49	morphosyntactic annotation part-of-speech annotation morphological annotation
7	inter-annotator kappa statistic inter-annotation agreementintra-annotator agreement inter-annotator agreement inter-coder agreement annotator agreement inter-annotator agreement scores inter-annotator agreements kappa coefficient inter-annotator agreement inter-rater agreement	18	morphosyntactic tags part-of-speech tag part-ofspeech tags part-of-speech part-of-speech tags partof-speech tags morphological tag partof-speech tag morphological tags part-of-speech labels morphological attributes part of speech parts-of-speech pos tags

subsets by time order. Each subset keeps 100 articles in average and represents a specific historical period's research overview.

Tracking Concept Popularities. In order to train the prediction model, we need to find out the actual popularity-change information of the concept as ground truth. Counting the frequency of concepts as their popularity is a common method in this area of research. Therefore, we count the popularity of the concept in each period and calculate the $trend(c, t)$ as trend of the concept. $trend(c, t)$ is defined as a three-valued quantity of 0,1,2. reflecting the state that $popularity(c, t_i)$ is equal to, larger than, and smaller than $popularity(c, t_{i+1})$. And $popularity(c, t)$ can be calculated as below.

$$popularity(c, t) = PaperwithConcpet(c) / SumOfPaper(t) \qquad (4)$$

Start the Rhetorical Predict Model. We use ArgZoneTagger to give each sentence in abstracts with 7 labels. Matched to the tagged abstracts, each concept presents a LDP feature in every period as Sect. 3.2 shows. Simultaneously, LDR and LDS feature can be calculated, so we finally find 1,676,976 terms of data for each feature. After pairing with the $trend$ data, we randomly select 75% of them as training data and the remaining 25% as test data, using L2 logistic regression(C=10) to construct the prediction model. We also train a topic model on same dataset of ACL like Vinodkumar's prior work to give a comparison. The results of predict model are shown in Table 5. The results show that using LDS feature gives best accuracy, while concepts instead of topics perform more validities of predicting the rise and fall of scientific trends, due to a better granularity and semantic concentration.

Table 5. The result of predict model using rhetorical features.

Feature of concept	Accuracy	Feature of topic	Accuracy
LDP	62.1%	LDP	60.3%
LDR	70.1%	LDR	70.8%
LDS	74.3%	LDS	72.0%

Using Related Concepts to Optimize the Model. Since the similarity of two concepts $sim(c_i, c_j)$ can be calculated, we select a concept's nearest related concepts' *trend* information in the former period as an assistant. We combine different number of related concepts and the accuracy changes like Table 6.

Table 6. The results of combined features

Test ID	Selected features	Accuracy
1	Nearest 7 concepts' recent trends	53.12%
2	LDP + Nearest 7 concepts' recent trends	66.44%
3	LDR + Nearest 7 concepts' recent trends	72.32%
4	LDS + Nearest 7 concepts' recent trends	81.21%
5	LDS + Nearest 8 concepts' recent trends	81.17%
6	LDS + Nearest 10 concepts' recent trends	80.10%
7	LDS + Nearest 14 concepts' recent trends	77.15%
8	LDS + Nearest 4 concepts' recent trends	80.82%
9	LDS + Nearest 6 concepts' recent trends	81.06%

The result indicates the help of related concepts and the difference between the numbers of related concepts involved. Test 1 tells the related concepts' information can only be an assistant. Test 2–4 tells LDP, LDR and LDS's accuracy is also like the results without new features. Test 4–10 shows that 7 related concepts seem to be the best to give the model an optimization, resulting in a 9% rise of accuracy compared with prior predict algorithm.

5   Conclusion and Future Work

In this paper, we propose a novel method to predict the rise-and-fall trends from the scientific concepts' rhetorical features. First we extract concepts from scientific literature and merge the identical concept mentions as research objects, then use ArgZoneTagger tool tagging sentences with 7 labels to discover rhetorical role of the concept in them. We calculate a concept's LDP, LDR, LDS feature to build the predict model and also consider the information of related concepts to optimize it. In experiments, we run the concept extractor in ACL1965-2017 dataset, resulting in 13,002 key phrases and 4,557 concept clusters. Besides, test shows changing the objects to concepts instead of topic gives a rise of predicting accuracy, while the model can perform better to with related concepts' recent trends.

Considering further, we can use the concepts dataset and their rhetorical features to analyze and discover more clues of scientific trends using other algorithm like RNN, which we leave as our future directions.

Acknowledgement. The work is supported by National Key Research and Development Program of China (2017YFB1002101), NSFC key project (U1736204, 61661146007) and THUNUS NExT Co-Lab.

References

1. Liu, Z., Huang, W., Zheng, Y., Sun, M.: Automatic keyphrase extraction via topic decomposition. In: Conference on Empirical Methods in Natural Language Processing, pp. 366–376, Association for Computational Linguistics (2010)
2. Griffiths, T.L., Steyvers, M.: Finding scientific topics. Proc. Natl. Acad. Sci. USA **101**(Suppl. 1), 5228–5235 (2004)
3. Anderson, A., Dan, M.F., Dan, J.: Towards a Computational History of the ACL: 1980–2008. In: ACL-2012 Special Workshop on Rediscovering 50 Years of Discoveries, pp. 13–21 (2013)
4. Blei, D.M., Lafferty, J.D.: Dynamic topic models. In: Proceedings of the International Conference on Machine Learning, pp. 113–120 (2006)
5. Blei, D.M., Ng, A.Y., Jordan, M.I.: Latent dirichlet allocation. J Mach. Learn. Res. Arch. **3**, 993–1022 (2003)
6. Hall, D., Jurafsky, D., Manning, C.D.: Studying the history of ideas using topic models. In: Proceedings of the Conference on Empirical Methods in Natural Language Processing, EMNLP 2008, 25–27 October 2008, Honolulu, Hawaii, USA, A Meeting of Sigdat, A Special Interest Group of the ACL, pp. 363–371. DBLP (2008)
7. Shibata, N., Kajikawa, Y., Takeda, Y., Matsushima, K.: Detecting emerging research fronts based on topological measures in citation networks of scientific publications. Technovation **28**(11), 758–775 (2008)
8. Shibata, N., Kajikawa, Y., Takeda, Y., Matsushima, K.: Comparative study on methods of detecting research fronts using different types of citation. J. Assoc. Inf. Sci. Technol. **60**(3), 571–580 (2009)
9. Mane, K.K., Börner, K.: Mapping topics and topic bursts in PNAS. Proc. Natl. Acad. Sci. USA **101**(Suppl 1), 5287–5290 (2004)

10. Guo, H., Weingart, S., Brner, K.: Mixed-indicators model for identifying emerging research areas. Scientometrics **89**(1), 421–435 (2011)
11. Small, H.: Tracking and predicting growth areas in science. Scientometrics **68**(3), 595–610 (2006)
12. Small, H.: Interpreting maps of science using citation context sentiments: a preliminary investigation. Scientometrics **87**(2), 373–388 (2011)
13. Prabhakaran, V., Hamilton, W.L., Dan, M.F., Dan, J.: Predicting the Rise and Fall of Scientific Topics from Trends in their Rhetorical Framing. In: Meeting of the Association for Computational Linguistics, pp. 1170–1180 (2016)
14. Grineva, M., Grinev, M., Lizorkin, D.: Extracting key terms from noisy and multi-theme documents. In: International Conference on World Wide Web, WWW 2009, Madrid, Spain, April, pp. 661–670. DBLP (2009)
15. Litvak, M., Last, M.: Graph-based keyword extraction for single-document summarization. In: MMIES 08 Workshop on Multi-source Multilingual Information Extraction & Summar, vol. 64, pp. 17–24 (2008)
16. Liu, Z., Li, P., Zheng, Y., Sun, M.: Clustering to find exemplar terms for keyphrase extraction. In: Conference on Empirical Methods in Natural Language Processing, vol. 1, PP. 257–266 (2009)
17. Turney, P.D.: Learning algorithms for keyphrase extraction. Inf. Retrieval **2**(4), 303–336 (2000)
18. Wan, X., Xiao, J.: Single document keyphrase extraction using neighborhood knowledge. In: National Conference on Artificial Intelligence, pp. 855–860. AAAI Press (2008)
19. Yuan, J., Gao, F., Ho, Q., Dai, W., Wei, J., Zheng, X., et al.: LightLDA: big topic models on modest computer clusters. 1351–1361 (2014)
20. Wang, X., Mccallum, A.: Topics over time: a non-Markov continuous-time model of topical trends. In: ACM SIGKDD International Conference on Knowledge Discovery and Data Mining, pp. 424-433. ACM (2006)
21. Turney, P.D.: Learning to extract keyphrases from text. cs.lg/0212013(cs.LG/0212013) (2002)
22. Mikolov, T., Sutskever, I., Chen, K., Corrado, G., Dean, J.: Distributed representations of words and phrases and their compositionality. Adv. Neural Inf. Process. Syst. **26**, 3111–3119 (2013)
23. Teufel, S.: Argumentative zoning: information extraction from scientific text (1999)
24. Liakata, M.: Zones of conceptualisation in scientific papers: a window to negative and speculative statements. In: The Workshop on Negation and Speculation in Natural Language Processing, pp. 1-4. Association for Computational Linguistics (2010)
25. Nguyen, T.D., Kan, M.-Y.: Keyphrase extraction in scientific publications. In: Goh, D.H.-L., Cao, T.H., Sølvberg, I.T., Rasmussen, E. (eds.) ICADL 2007. LNCS, vol. 4822, pp. 317–326. Springer, Heidelberg (2007). https://doi.org/10.1007/978-3-540-77094-7_41
26. Mihalcea, R.: Textrank: bringing order into texts. In: EMNLP, pp. 404–411 (2004)

Knowledge Augmented Inference
Network for Natural Language Inference

Shan Jiang[2], Bohan Li[2], Chunhua Liu[2], and Dong Yu[1,2(✉)]

[1] Beijing Advanced Innovation for Language Resources of BLCU, Beijing, China
yudong_blcu@126.com
[2] Beijing Language and Culture University, Beijing, China
jiangshan727@gmail.com, bohanli.lavida@gmail.com, chunhualiu596@gmail.com

Abstract. This paper proposes a Knowledge Augmented Inference Network (K- AIN) that can effectively incorporate external knowledge into existing neural network models on Natural Language Inference (NLI) task. Different from previous works that use one-hot representations to describe external knowledge, we employ the TransE model to encode various semantic relations extracted from the external Knowledge Base (KB) as distributed relation features. We utilize these distributed relation features to construct knowledge augmented word embeddings and integrate them into the current neural network models. Experimental results show that our model achieves a better performance than the strong baseline on the SNLI dataset and we also surpass the current state-of-the-art models on the SciTail dataset.

Keywords: Natural language inference · External knowledge
Knowledge graph embedding

1 Introduction

Natural Language Inference (NLI) is a task that determines whether a hypothesis sentence can be inferred from a premise sentence, which is an essential and challenging task for Natural Language Understanding (NLU) [11]. Due to the availability of large annotated datasets such as SNLI [2] and MultiNLI [27], neural network models have made remarkable headway.

There are two main categories of the neural network models on NLI: one is single sentence encoder-based models which focus on sentence encoding and representation [2,4,14,20], and the other is cross sentence attention-based models which concern more about the interaction between a sentence pair [3,15,18,25,26]. Among these models, ESIM [3] is a competitive model that enhances sequential inference information based on chain networks. Many studies are motivated by this model [6,23]. All above models are trained end to end and all their inference information is learned from static training datasets. Intuitively, these limited datasets can not cover all the necessary semantic inference

© Springer Nature Singapore Pte Ltd. 2019
J. Zhao et al. (Eds.): CCKS 2018, CCIS 957, pp. 129–135, 2019.
https://doi.org/10.1007/978-981-13-3146-6_11

information. So we naturally explore an approach that leverages the external semantic knowledge to enrich the current neural NLI models.

External knowledge has been proved to be helpful in various natural language processing tasks, such as machine translation [21], machine comprehension [10] and dialogue response generation [24]. However, there are only a few works exploring external knowledge on NLI. [19] proposes MKAL model that transfers the original training sentences into a series of triplets and utilizes CNN and PtransE to calculate the semantic embeddings. [5] extracts knowledge of free text and dynamically incorporates them to refine word embeddings. [17] integrates knowledge of binary representations to enrich the ESIM model. The main difference between our model and the previous works is that we represent external knowledge in the form of distributed and structured triplets and then leverage them to argument word embedding information.

In this paper, we propose a Knowledge Augmented Inference Network (KAIN) that equips the current neural NLI models with distributed knowledge features. We achieve better performance than the strong baseline model ESIM on the SNLI dataset. Furthermore, we obtain an accuracy of 85.0% on the SciTail dataset which is the new state-of-the-art result.

Table 1. Semantic relations extracted from WordNet.

Relations	#Pairs	Triplet examples
Synonym	12254	(young, Synonym, youth)
Antonym	10828	(young, Antonym, age)
Hypernym	165769	(young, Hypernym, creature)
Hyponym	114729	(people, Hyponym, young)
Same Hpernym	56939	(young, Same Hpernym, adult)

2   Model

2.1   Knowledge Extraction and Representation

In this paper, we choose WordNet [13] as our external knowledge resource. Inspired by [17], we extract various semantic relations including Synonym, Antonym, Hypernym, Hyponym and Same Hypernym, then further construct the knowledge triplets. The statistical details and examples are shown in Table 1. Then we employ TransE [1], a basic KB embedding approach, to calculate the embeddings of each entity and relation in the triplets. The relation embeddings are regarded as inputs fed into the Knowledge Augmented Embedding Layer of KAIN.

2.2 Knowledge Augmented Inference Network

Knowledge Augmented Embedding Layer. This is the core layer within our model that utilizes the pre-trained knowledge triplets to construct the knowledge augmented word representations. Let $P = [p_1, p_2, \cdots, p_M]$ and $Q = [q_1, q_2, \cdots, q_N]$ denote the original premise and hypothesis of M words and N words respectively. Here we assume that each p_i ($1 \leq i \leq M$)and q_j ($1 \leq j \leq N$) are standard word embeddings of d dimensions. For the retrieved semantic knowledge from KB, we construct and embed knowledge triplets into (p_i, r_{ij}, q_j), where $r_{ij} \in \mathcal{R}^e$. We will describe how to calculate the knowledge-refined premise $\bar{P}$, and the same approach is applied to obtain $\bar{Q}$ (Fig. 1).

For each premise word p_i, we first process the corresponding knowledge triplet (p_i, r_{ij}, q_j) by two non-linear map functions as follows:

$$f^p(r_{ij}) = Relu(W_s r_{ij} + b_s) \tag{1}$$

$$f^p(q_j) = Relu(W_v q_j + b_v), \tag{2}$$

where W_s, b_s, W_v, and b_v are weights to be learned. Then we use a dot product function to integrate $f^p(r_{ij})$ and $f^p(q_j)$, and add them to the original premise embedding vector p_i as follows:

$$\bar{p}_i = p_i + \sum_{j}^{N} f^p(r_{ij}) \odot f^p(q_j), \tag{3}$$

where $\bar{p}_i \in \mathcal{R}^d$ is a knowledge augmented premise word embedding. It contains not only the original semantic information of p_i but the related word and relation information from the whole Q.

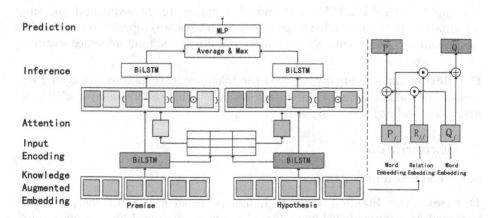

Fig. 1. Architecture of KAIN and the datails about the Knowledge Augment Emebedding Layer.

Input Encoding Layer. This layer aims to learn the contextual information about premise and hypothesis. We utilize a Bi-directional LSTM [7] (BiLSTM) to encode the knowledge augmented sequences $\bar{P}$ and $\bar{Q}$ at each time-step, then obtain the hidden states $H^P \in \mathcal{R}^{M \times 2d}$ and $H^Q \in \mathcal{R}^{N \times 2d}$.

Attention Layer. Here we apply a soft alignment mechanism to attend and further compose the relevant components of the sequence pair. We first calculate the similarity matrix by $A = (H^P)^T H^Q \in \mathcal{R}^{M \times N}$. We denote $softmax(M)$ as the softmax operation over the matrix M which normalizes M by column-wise. We align and calculate the weighted summaries of the premise and hypothesis as:

$$\widetilde{P} = softmax(A)H^Q \qquad\qquad \widetilde{Q} = softmax(A^T)H^P \qquad (4)$$

where $\widetilde{P} \in \mathcal{R}^{M \times R^{2d}}$ is the extracted relevant features of H^Q by attending to H^P and vice versa for $\widetilde{Q} \in \mathcal{R}^{N \times R^{2d}}$.

Inference Layer. This layer is to further integrate the collected attentional information. Inspired by [3], we apply the difference and the element-wise product to the original vector tuples $(\widetilde{P}, H^p)$ and $(\widetilde{Q}, H^Q)$. This provides the enhanced difference and similar features. Next, we concatenate all features by Eqs. 5 and 6 and project them into d dimensional vector space by a feed-forward neural network with *ReLU* activation respectively, which constructs the high-level interaction features $\hat{P}$ and $\hat{Q}$.

$$\hat{P} = F[\widetilde{P}, H^P, \widetilde{P} - H^P, \widetilde{P} \odot H^P] \qquad (5)$$

$$\hat{Q} = F[\widetilde{Q}, H^Q, \widetilde{Q} - H^Q, \widetilde{Q} \odot H^Q] \qquad (6)$$

We apply another BiLSTM to $\hat{P}$ and $\hat{Q}$ to aggregate the sequential inference information. Both max and average pooling methods are employed on the hidden states and we concatenate all the pooling results as the final inference features.

Prediction Layer. Finally, we feed the inference features into two-layer feed-forward neural networks for classification which include a hidden layer with *tanh* activation and *softmax* output layer.

3   Experiment

3.1   Experiment Setting

Dataset. The Stanford Natural Language Inference (SNLI) corpus [2] is regarded as a major benchmark for NLI, which contains 570k sentence pairs. Each annotated sentence pair in SNLI is consist of a premise, a hypothesis, and a golden label (*entailment, contradiction,* or *neutral*). The SciTail [8] is a small scale dataset that contains 24k sentence pairs and the golden label is *entailment* and *neural*.

Table 2. Performance on SNLI

Models	Train(%acc)	Test(%acc)
LSTM Encoders [2]	84.8	77.6
WbW Attention [18]	85.3	83.5
mLSTM [25]	92.0	86.1
ESIM [3]	92.6	**88.0**
KIM [17]	94.1	88.6
KAIN	93.7	**88.2**
KAIN (Ensemble)	-	**88.8**

Table 3. Performance on SciTail

Models	Valid(%acc)	Test(%acc)
Majority class*	63.3	60.3
DecompAtt*	75.4	72.3
ESIM*	70.5	**70.6**
Ngram*	65.0	70.6
DGEM*	79.6	77.3
CAFE [23]	-	83.3
KAIN	85.3	**85.0**

Training. The word embeddings are initialized by $300D$ Glove $840B$ [16] vectors and out-of-vocabulary words are initialized randomly. The dimensions of the relation and the hidden states of BiLSTMs are $20D$ and $300D$ respectively. All weights are constrained by L2 regularization. We use dropout [22] to all the layers with a dropout rate of 0.2. The model is optimized with Adam [9] with an initial learning rate of 0.0003. Before extract external knowledge, we lemmatize the training data by Stanford CoreNLP [12]. To help duplicate our model, we will release our source code at [xxx].

3.2 Results and Analysis

Table 2 shows the accuracies of our KAIN model along with the current published models on SNLI training and test set. [3] introduces ESIM, the sequential inference model with local inference, which achieves 88.0% accuracy. In this paper, we choose ESIM as our baseline and utilize our KAIN model to equip it. We surpass the ESIM by 0.2% which proves the current neural network model can benefit from the external knowledge. By leveraging the embedded knowledge triplets, our model constructs the knowledge augmented word embeddings, which capture not only the original single-word information but also the related word and relation information. The supplementary knowledge information flows through the neural networks to finally assist in reasoning.

Table 3 demonstrates the experimental results on SciTail valid and test set. Models with * are reported from [8]. Our KAIN model achieves 85.0% in accuracy which outperforms all the previous results. Compared with the baseline model ESIM and the current state-of-the-art model CAFE [23], we yield 14.4% and 1.7% improvements respectively, which again proves the effectiveness of our model. Furthermore, we observe the Tables 2 and 3 simultaneously and find that external knowledge provides more assistance to models on small datasets than large datasets. Intuitively, neural network models on large datasets have strong learning ability that is able to capture abundant information for reasoning. In contrary, small datasets can't provide sufficient necessary inference information and our model exactly utilizes external knowledge to help neural network models fill this gap.

4   Conclusion

In this paper, we introduce the Knowledge Augmented Inference Network (KAIN) that leverages external knowledge to enrich the inference information for the current neural NLI models. We surpass the strong baseline ESIM on the SNLI dataset and achieve a new state-of-the-art accuracy on the SciTail dataset. Experimental results prove that the neural NLI models can benefit from the distributed external knowledge. For future work, we will keep exploring which kind of knowledge can assist neural network models reason better and how to represent and leverage them more effectively.

Acknowledgements. This work is funded by Beijing Advanced Innovation for Language Resources of BLCU, the Fundamental Research Funds for the Central Universities in BLCU (No.17PT05) and the BLCU Academic Talents Support Program for the Young and Middle-Aged.

References

1. Bordes, A., Usunier, N., García-Durán, A., Weston, J., Yakhnenko, O.: Translating embeddings for modeling multi-relational data. In: NIPS (2013)
2. Bowman, S.R., Angeli, G., Potts, C., Manning, C.D.: A large annotated corpus for learning natural language inference. In: ACL (2015)
3. Chen, Q., Zhu, X., Ling, Z.H., Wei, S., Jiang, H., Inkpen, D.: Enhanced LSTM for natural language inference. In: ACL (2017)
4. Choi, J., Yoo, K.M., Lee, S-G.: Unsupervised learning of task-specific tree structures with tree-LSTMs. CoRR abs/1707.02786 (2017)
5. Weissenborn, D., Kočiský, T., Dyer, C.: Reading twice for natural language understanding. CoRR (2017)
6. Ghaeini, R., et al.: DR-BiLSTM: dependent reading bidirectional LSTM for natural language inference. CoRR (2018)
7. Hochreiter, S., Schmidhuber, J.: Long short-term memory. Neural Comput. **9**, 1735–1780 (1997)
8. Khot, T., Sabharwal, A., Clark, P.: SciTail: a textual entailment dataset from science question answering (2018)
9. Kingma, D.P., Ba, J.: Adam: a method for stochastic optimization. CoRR (2014)
10. Lin, H., Sun, L., Han, X.: Reasoning with heterogeneous knowledge for commonsense machine comprehension (2017)
11. MacCartney, B., Manning, C.D.: Modeling semantic containment and exclusion in natural language inference. COLING (2008)
12. Manning, C., Surdeanu, M., Bauer, J., Finkel, J., Bethard, S., McClosky, D.: The Stanford CoreNLP natural language processing toolkit. In: ACL (2014)
13. Miller, G.A.: WordNet: a lexical database for English. Commun. ACM **38**, 39–41 (1992)
14. Nie, Y., Bansal, M.: Shortcut-stacked sentence encoders for multi-domain inference. In: ACL (2017)
15. Parikh, A.P., Täckström, O., Das, D., Uszkoreit, J.: A decomposable attention model for natural language inference. CoRR (2016)
16. Pennington, J., Socher, R., Manning, C.D.: Glove: global vectors for word representation. In: EMNLP (2014)

17. Chen, Q., Zhu, X., Ling, Z.-H., Inkpen, D., Wei, S.: Natural language inference with external knowledge. CoRR (2017)
18. Rocktäschel, T., Grefenstette, E., Hermann, K.M., Kociský, T., Blunsom, P.: Reasoning about entailment with neural attention. CoRR (2015)
19. Sha, L., Li, S., Chang, B., Sui, Z.: Recognizing textual entailment via multi-task knowledge assisted LSTM (2016)
20. Shen, T., Zhou, T., Long, G., Jiang, J., Wang, S., Zhang, C.: Reinforced self-attention network: a hybrid of hard and soft attention for sequence modeling. CoRR abs/1801.10296 (2018)
21. Shi, C., et al.: Knowledge-based semantic embedding for machine translation (2016)
22. Srivastava, N., Hinton, G., Krizhevsky, A., Sutskever, I., Salakhutdinov, R.: Dropout: a simple way to prevent neural networks from overfitting. In: Machine Learning Research, pp. 1929–1958 (2014)
23. Tay, Y., Tuan, L.A., Hui, S.C.: A compare-propagate architecture with alignment factorization for natural language inference. CoRR (2018)
24. Vougiouklis, P., Hare, J.S., Simperl, E.P.B.: A neural network approach for knowledge-driven response generation. In: COLING (2016)
25. Wang, S., Jiang, J.: Learning natural language inference with LSTM. CoRR (2015)
26. Wang, Z., Hamza, W., Florian, R.: Bilateral multi-perspective matching for natural language sentences. CoRR (2017)
27. Williams, A., Nangia, N., Bowman, S.R.: A broad-coverage challenge corpus for sentence understanding through inference. CoRR (2017)

Survey on Schema Induction
from Knowledge Graphs

Qiu Ji[1]([✉]), Guilin Qi[2], Huan Gao[2], and Tianxing Wu[2]

[1] School of Modern Posts and Institute of Modern Posts,
Nanjing University of Posts and Telecommunications, Nanjing, China
qiuji@njupt.edu.cn
[2] School of Computer Science and Engineering, Southeast University, Nanjing, China
{gqi,gh,wutianxing}@seu.edu.cn

Abstract. With the rapid growth of knowledge graphs, schema induction, as a task of extracting relations or constraints from a knowledge graph for the classes and properties, becomes more critical and urgent. Schema induction plays an important role to facilitate many applications like integrating, querying and maintaining knowledge graphs. To provide a comprehensive survey of schema induction, in this paper, we overview existing schema induction approaches by mainly considering their learning methods, the types of learned axioms and the external resources that may be used during the learning process. Based on the comparison, we point out the challenges and directions for schema induction.

Keywords: Ontology learning · Schema induction
Knowledge graph · Semantic web

1 Introduction

With the effort of linked open data and the systems to automatically generate semantic data, a large number of knowledge graphs (KGs) become available, such as DBpedia and NELL. However, such KGs mainly contain instance-level facts while lacking schema information like disjointness and subsumptions. This make it difficult to integrate, query and maintain KGs [21]. To deal with this problem, schema induction is performed to extract schema information from a KG. Since an ontology consists of a TBox of terminology axioms and an ABox of individual axioms, schema induction actually is to learn TBox axioms.

Currently, many diverse solutions to the schema induction problem have been proposed, e.g., [4,5,8,21], and good surveys are given. The survey of [3] focuses on the state of the art in ontology learning from text. The work in [16] summaries those machine learning techniques that are used to solve ontology mining problems. The survey given in [17] reviews most of the recent schema learning approaches. However, these surveys either do not provide clear and comprehend classifications to facilitate users for choosing an appropriate schema induction approach, or many recently proposed approaches are not included.

© Springer Nature Singapore Pte Ltd. 2019
J. Zhao et al. (Eds.): CCKS 2018, CCIS 957, pp. 136–142, 2019.
https://doi.org/10.1007/978-981-13-3146-6_12

For the reasons above, in this paper, we provide a thorough overview of schema induction from KGs according to different classifications. Specifically, we mainly consider the learning methods, the types of learned axioms and the external resources that may be used during the learning process. Based on the comparison, we point out the challenges and directions for schema induction.

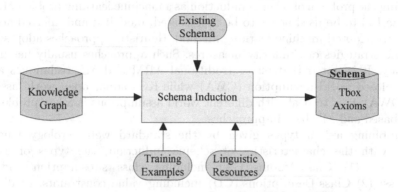

Fig. 1. The process of schema induction from knowledge graphs.

2   Problem Statement

Generally, schema induction takes a KG as an input and outputs a schema consisting of the learned TBox axioms with or without external resources like linguistic resources and existing schemas (see Fig. 1). Usually, a relevant existing schema could be reused as constraints to enhance the learning effectiveness. If a supervised learning approach is applied, training examples are required. Furthermore, the linguistic resources like thesauri may be used to find more relations among the individuals, properties or classes.

3   Classifications of Schema Induction Approaches

To thoroughly review the existing approaches about schema induction, we classify existing schema induction approaches by mainly considering their learning methods, the types of learned axioms and the external resources to be used. Since it is obvious to use external resources to distinguish different schema induction approaches, we focus on the introduction of the former two classifications.

According to the learning methods, schema induction approaches can be roughly divided into inductive logic programming (ILP)-based approaches, association rule mining (ARM)-based approaches, machine learning (ML)-based approaches and heuristic approaches. ILP-based approaches combine machine learning techniques with logic programming to construct schemas from instances and background knowledge. During the learning process, a new axiom is regarded as correct if the testing KG plus it can infer the positive examples and not infer the negative ones. ARM-based approaches usually construct transaction tables based on the information extracted from a KG first, and then apply a mining method to find association rules. Such rules often could be transferred to TBox axioms easily. ML-based approaches use various machine learning methods by regarding the problem of schema induction as a machine learning problem. In this way, the KG to be used needs to be represented, modelled and inferred according to the adopted machine learning method. Heuristic approaches adopts some heuristic strategies or similarity measures. Such approaches usually has higher efficiency. It is noted that, since traditional ARM and ML techniques adopt (local) closed world assumption (CWA) while KGs adopt open world assumption (OWA), how to deal with different world assumptions is a key problem for ARM-based and ML-based approaches.

Combining axiom types given by the standard web ontology language OWL[1] with the characteristics of schema induction, the types of axioms consist of: (1) Class Axioms (CA) including class subsumption and disjointness; (2) Class Descriptions (CD) including value constraints, cardinality constraints and set operations[2]; (3) Property Axioms (PA) including property subsumption, disjointness, domain, range and inverse properties; (4) Property Characteristics (PC) including functionality, symmetry and transitivity. For simplicity, we use the first three letters to indicate each category. For example, val indicates value constraints. Especially, we use sub to represent both class and property subsumption, and use c-sub to indicate class subsumption. Other categories can be abbreviated similarly.

4   Analysis of Schema Induction Approaches

With the classifications given in Sect. 3, we give a comprehensive analysis of the existing works about schema induction. See Table 1, where "–" indicates no external resources are used and "n.a." means not available currently.

At the early age of schema induction research, the ILP-based approach [11] was proposed by using refinement operators to learn class descriptions from instance data and background ontology. Later on, this work was extended in [9] to deal with larger KGs by proposing strategies to select "interesting parts". Another extension is given in [12], which can deal with more expressive KGs by learning more types of axioms such as cardinality restrictions. The corresponding methods were implemented in DL-Learner with a graphical user interface. These methods have been shown effective, but high quality of datasets are required and the scalability is still a big problem.

---

[1] https://www.w3.org/TR/owl-ref/.
[2] Set operations mean the intersection, union or negation of classes.

Table 1. The comparison of existing schema induction works.

Learning methods	Existing works	Axiom types	External resources	Tools
ILP-based	Lehmann07 [11]	val, set, CA	TraExa, ExiSch	DL-Learner[a]
	Lehmann10 [12]	CA, CD	TraExa, ExiSch	DL-Learner
ARM-based	Voelker11 [21]	c-dis, sub, dom, ran, val	–	GoldMiner[b]
	Fleischhacker11 [5]	c-dis	–	GoldMiner
	Fleischhacker12 [6]	PA, PC	–	GoldMiner
	Ell16 [4]	dom, ran	–	n.a.
	Irny17 [10]	inv, s8m	–	n.a.
	Gao18 [8]	c-dis, c-sub	–	SIFS[c]
ML-based	Voelker07 [19]	c-dis	TraExa,ExiSch,LinRes	LeDA[d]
	Zhu14 [22]	c-dis, c-sub	TraExa	n.a.
	Rizzo17 [15]	c-dis	ExiSch	TCT[e]
Heuristic	Meilicke08 [13]	c-dis	–	n.a.
	Gerald12 [18]	c-dis, dom, ran	–	n.a.
	Buehmann12 [1]	CA, PA, PC	–	DL-Learner
	Buehmann13 [2]	CA, PA, PC, val, set	–	DL-Learner
	Munoz17 [14]	car	–	n.a

[a]http://dl-learner.org/
[b]https://github.com/dfleischhacker/goldminer/tree/reorganization
[c]https://github.com/gaohuan2015/SIFS
[d]https://code.google.com/p/leda-project/
[e]http://github.com/Giuseppe-Rizzo/TCT

For ARM-based approaches[3], the first approach was given in [21], which was extended by [5] and [6] to learn disjointness and property axioms separately by defining various association rule patterns. It was also extended by [20] to generate negative association rules for learning disjointness. Besides, by modifying some technical details in [21], the work in [4] can induce not only independent domain and range restrictions but also coupled ones. Since these methods always adopt CWA which is opposite to the assumption adopted by KGs, a lot of noisy negative examples may be generated which may lead to poor performance. Thus, the work in [8] proposed a novel framework to obtain probabilistic type assertions by exploiting a type inference algorithm and then redefined the mining model accordingly. Besides, the work in [10] used ARM to learn inverse and symmetric axioms with new measures to rank the rules of interest. It also proposed an unsupervised method to determine the confidence threshold for identifying interesting rules.

ML-based approaches try different machine learning models to learn schemas. The authors in [19] adopted the ADTree classifier to learn disjointness axioms. They used various external resources like manually constructed training examples, background ontologies, textual resources and WordNet to obtain features. This work was extended in [20] to use more logical features and external resources like wikipedia. The work in [22] integrated the probabilistic inference capability of Bayesian Networks with the logical formalism of Description Logics. To be consistent with OWA of KGs, the traditional confusion matrix was extended

---

[3] Although some existing works like [7] could generate association rules from KGs, we do not include them in this paper since their goal is not to generate schemas.

to consider unknown results. The work in [15] considered the task of discovering disjointness axioms as a clustering problem by using the terminological cluster trees. This method combined distance-based clustering and the divide-and-conquer strategy.

As for heuristic approaches, the work in [13] assumed that all sibling concepts are disjoint. In [18], property domain and range can be obtained by defining two metrics with statistical methods over DBpedia. Besides, when generating class disjointness, all concepts were mapped to vectors with the same dimension and two concepts were regarded as disjoint if their similarity is lower than a threshold. A light-weight approach was presented in [1], which used SPARQL patterns to generate nearly all kinds of TBox axioms except class descriptions. Thus, their subsequent work [2] detected frequent axiom usage patterns from more than 1000 ontologies and then converted them to SPARQL patterns for learning almost all types of TBox axioms. The work in [14] used the semantics of owl:sameAs equality axioms in KGs to generate accurate cardinalities.

5   Challenges and Directions

Although various approaches have been proposed to learn schemas from KGs, schema induction is still a challenging research topic to achieve better performance and user experience.

Since KGs often own large amount of instance data which lack of balance and keep evolving, existing schema induction approaches may not be good enough to cope with. Thus, as a future work, linguistic resources like WordNet can be used to enrich instance data for coping with the imbalance problem. Besides, we could use an iterative learning process with existing schemas as constraints to take the dynamics into account. To obtain better performance, it would be good to integrate ARM-based approaches and heuristic ones.

To achieve better user experience, a friendly graphical user interface is urgently required. As we can see from the existing works, although the proposed schema induction approaches may have been implemented, the corresponding APIs or tools are not available online for some reasons. Currently, very few schema induction tools are available and only DL-Learner provides a user interface. For future development, a plugin-based framework is desired by providing various learning methods to learn user-specified types of axioms.

6   Conclusions

This paper systematically presented the problem of schema induction and classified schema induction approaches from multiple perspectives. The first classification was made based on the learning methods. The second one is made by combining axiom types given by the standard web ontology language OWL with the characteristics of schema induction. The third one is made by external resources. With these classifications, we thoroughly reviewed the existing schema induction approaches. We observed that, ARM and heuristic strategies are more welcome due to their high efficiency and external resources have not been well exploited. Finally, we pointed out the challenges and future directions.

Acknowledgements. This paper is sponsored by NSFC 61602259 and U1736204.

References

1. Bühmann, L., Lehmann, J.: Universal OWL axiom enrichment for large knowledge bases. In: ten Teije, A., et al. (eds.) EKAW 2012. LNCS (LNAI), vol. 7603, pp. 57–71. Springer, Heidelberg (2012). https://doi.org/10.1007/978-3-642-33876-2_8
2. Bühmann, L., Lehmann, J.: Pattern based knowledge base enrichment. In: Alani, H., et al. (eds.) ISWC 2013. LNCS, vol. 8218, pp. 33–48. Springer, Heidelberg (2013). https://doi.org/10.1007/978-3-642-41335-3_3
3. Cimiano, P.: Ontology Learning and Population from Text - Algorithms, Evaluation and Applications. Springer, New York (2006). https://doi.org/10.1007/978-0-387-39252-3
4. Ell, B., Hakimov, S., Cimiano, P.: Statistical induction of coupled domain/range restrictions from RDF knowledge bases. In: van Erp, M., Hellmann, S., McCrae, J.P., Chiarcos, C., Choi, K.-S. (eds.) ISWC 2016. LNCS, vol. 10579, pp. 27–40. Springer, Cham (2016). https://doi.org/10.1007/978-3-319-68723-0_3
5. Fleischhacker, D., Völker, J.: Inductive learning of disjointness axioms. In: Meersman, R., et al. (eds.) OTM 2011. LNCS, vol. 7045, pp. 680–697. Springer, Heidelberg (2011). https://doi.org/10.1007/978-3-642-25106-1_20
6. Fleischhacker, D., Völker, J., Stuckenschmidt, H.: Mining RDF data for property axioms. In: Meersman, R., et al. (eds.) OTM 2012. LNCS, vol. 7566, pp. 718–735. Springer, Heidelberg (2012). https://doi.org/10.1007/978-3-642-33615-7_18
7. Galarraga, L., Teflioudi, C., Hose, K., Suchanek, F.M.: Fast rule mining in ontological knowledge bases with AMIE+. VLDB J. **24**(6), 707–730 (2015)
8. Gao, H., Qi, G., Ji, Q.: Schema induction from incomplete semantic data. In: Intelligent Data Analysis (2018, to appear)
9. Hellmann, S., Lehmann, J., Auer, S., Sheth, A.: Learning of OWL class descriptions on very large knowledge bases. Int. J. Semant. Web Inf. Syst. **5**(2), 25–48 (2009)
10. Irny, R., Kumar, P.S.: Mining inverse and symmetric axioms in linked data. In: JIST, pp. 215–231 (2017)
11. Lehmann, J., Hitzler, P.: A refinement operator based learning algorithm for the $\mathcal{ALC}$ description logic. In: Blockeel, H., Ramon, J., Shavlik, J., Tadepalli, P. (eds.) ILP 2007. LNCS (LNAI), vol. 4894, pp. 147–160. Springer, Heidelberg (2008). https://doi.org/10.1007/978-3-540-78469-2_17
12. Lehmann, J., Hitzler, P.: Concept learning in description logics using refinement operators. Mach. Learn. **78**(1–2), 203–250 (2010)
13. Meilicke, C., Völker, J., Stuckenschmidt, H.: Learning disjointness for debugging mappings between lightweight ontologies. In: Gangemi, A., Euzenat, J. (eds.) EKAW 2008. LNCS (LNAI), vol. 5268, pp. 93–108. Springer, Heidelberg (2008). https://doi.org/10.1007/978-3-540-87696-0_11
14. Muñoz, E., Nickles, M.: Mining cardinalities from knowledge bases. In: Benslimane, D., Damiani, E., Grosky, W.I., Hameurlain, A., Sheth, A., Wagner, R.R. (eds.) DEXA 2017. LNCS, vol. 10438, pp. 447–462. Springer, Cham (2017). https://doi.org/10.1007/978-3-319-64468-4_34

15. Rizzo, G., d'Amato, C., Fanizzi, N., Esposito, F.: Terminological cluster trees for disjointness axiom discovery. In: Blomqvist, E., Maynard, D., Gangemi, A., Hoekstra, R., Hitzler, P., Hartig, O. (eds.) ESWC 2017. LNCS, vol. 10249, pp. 184–201. Springer, Cham (2017). https://doi.org/10.1007/978-3-319-58068-5_12

16. Sheu, P., Yu, H., Ramamoorthy, C.V., Joshi, A.K.: Machine Learning Methods for Ontology Mining. Wiley-IEEE Press, Hoboken (2010)

17. Subhashree, S., Irny, R., Sreenivasa Kumar, P.: Review of approaches for linked data ontology enrichment. In: Negi, A., Bhatnagar, R., Parida, L. (eds.) ICDCIT 2018. LNCS, vol. 10722, pp. 27–49. Springer, Cham (2018). https://doi.org/10.1007/978-3-319-72344-0_2

18. Toepper, G., Knuth, M., Sack, H.: DBpedia ontology enrichment for inconsistency detection. In: I-SEMANTICS, pp. 33–40 (2012)

19. Völker, J., Vrandečić, D., Sure, Y., Hotho, A.: Learning disjointness. In: Franconi, E., Kifer, M., May, W. (eds.) ESWC 2007. LNCS, vol. 4519, pp. 175–189. Springer, Heidelberg (2007). https://doi.org/10.1007/978-3-540-72667-8_14

20. Völker, J., Fleischhacker, D., Stuckenschmidt, H.: Automatic acquisition of class disjointness. J. Web Semant. **35**, 124–139 (2015)

21. Völker, J., Niepert, M.: Statistical schema induction. In: Antoniou, G., et al. (eds.) ESWC 2011. LNCS, vol. 6643, pp. 124–138. Springer, Heidelberg (2011). https://doi.org/10.1007/978-3-642-21034-1_9

22. Zhu, M., Gao, Z., Pan, J.Z., Zhao, Y., Ying, X., Quan, Z.: Tbox learning from incomplete data by inference in BelNet+. Knowl.-Based Syst. **75**(C), 30–40 (2014)

Author Index

Cao, Pengfei   91
Chen, Huajun   52
Chen, Jiaoyan   52
Chen, Yongrui   28
Chen, Yubo   91
Cheng, Gong   1
Cheng, Shuzhi   14

Dai, Xinyu   1
Deng, Shumin   52
Du, Xiaoping   116

Gao, Huan   136
Ge, Weiyi   65
Geng, Ji   78
Guo, Lingbing   65
Guo, Tong   78
Guo, Zehao   40

He, Weizhuo   1
Hu, Changjian   78
Hu, Wei   65

Ji, Qiu   136
Jiang, Shan   129

Li, Bohan   129
Li, Huiying   28
Li, Juanzi   116
Li, Xufeng   103
Li, Yang   78
Liu, Chunhua   129
Liu, Kang   91

Liu, Liting   14
Liu, Qizhi   1

Ma, Hongyuan   103
Miao, Qingliang   78

Pan, Jeff Z.   52
Pan, Liangming   116

Qi, Guilin   136
Qu, Yuzhong   1, 65

Shi, Wenxuan   14
Sun, Zequn   1

Wang, Zhenyu   40
Wu, Tianxing   136

Xu, Feiyu   78
Xu, Zejian   28

Yang, Jianlei   103
Yu, Dong   129
Yu, Jifan   116

Zhang, Hao   1
Zhang, Hualong   14
Zhang, Lingling   1
Zhang, Qingheng   65
Zhang, Richong   103
Zhang, Rui   40
Zhang, Zhiwei   1
Zhao, Jun   91

Printed in the United States
By Bookmasters